Academic English for Energy

能源学术英语 ②

An Integrated Course
综合教程

主　编：赵秀凤
副主编：陈　芳　徐方富
编　者：王文征　蔡　坤　单小明　连　洁　郑世高
　　　　李子男　田　芸　王　葳　李　音　石　虹

外语教学与研究出版社
FOREIGN LANGUAGE TEACHING AND RESEARCH PRESS
北京 BEIJING

图书在版编目 (CIP) 数据

能源学术英语综合教程. 2 / 赵秀凤主编；王文征等编. -- 北京：外语教学与研究出版社，2019.10（2023.7重印）
　　ISBN 978-7-5213-1254-6

Ⅰ. ①能… Ⅱ. ①赵… ②王… Ⅲ. ①能源－英语－教材 Ⅳ. ①TK

中国版本图书馆 CIP 数据核字 (2019) 第 239561 号

出 版 人　王　芳
项目负责　孟　琪
责任编辑　陈　玲
责任校对　毕海英
装帧设计　涂　俐
出版发行　外语教学与研究出版社
社　　址　北京市西三环北路 19 号（100089）
网　　址　https://www.fltrp.com
印　　刷　北京虎彩文化传播有限公司
开　　本　889×1194　1/16
印　　张　10.5
版　　次　2020 年 1 月第 1 版　2023 年 7 月第 5 次印刷
书　　号　ISBN 978-7-5213-1254-6
定　　价　49.80 元

如有图书采购需求，图书内容或印刷装订等问题，侵权、盗版书籍等线索，请拨打以下电话或关注官方服务号：
客服电话：400 898 7008
官方服务号：微信搜索并关注公众号"外研社官方服务号"
外研社购书网址：https://fltrp.tmall.com

物料号：312540101

"能源学术英语"系列教材
编委会

（以姓氏拼音为序）

白雪晴	东北石油大学
陈会军	中国地质大学（北京）
高　霄	华北电力大学（保定）
韩淑芹	中国石油大学（华东）
李特夫	广东石油化工学院
梁建民	辽宁石油化工大学
吕旭英	西南石油大学
苗亚男	东北电力大学
潘卫民	上海电力大学
谈宏慧	长江大学
吴格非	中国矿业大学
许卉艳	中国矿业大学（北京）
严娇兰	北京石油化工学院
于艳英	西安石油大学
张峻峰	中国地质大学（武汉）
赵秀凤	中国石油大学（北京）
赵玉闪	华北电力大学

前 言

能源是工业文明的基础，是人类文明进步的动力，塑造着人类的历史、现状和未来。在全球化的今天，能源问题已经成为国际政治、经济、环境保护等诸多领域的中心议题。当今国际社会的焦点议题如气候变化、低碳经济、可持续发展等无不与能源相关——这些话题不仅牵涉到国际格局和国家发展，也与民众的生活息息相关。

党的二十大报告指出，尊重自然、顺应自然、保护自然，是全面建设社会主义现代化国家的内在要求。我们要深入推进能源革命，加快推进能源结构调整优化，践行绿水青山就是金山银山的理念，站在人与自然和谐共生的高度谋划发展。因此，通过学习与能源相关的热点话题，审视人类文明和国际国内社会生活，培养大学生的国际视野和家国情怀，为推进美丽中国建设培育新一代人才，是《能源学术英语综合教程》的立意所在。

本教材涉及能源与日常生活、能源与技术、能源与伦理、能源与可持续发展、能源与安全、能源与地缘政治变化、能源与文明、能源与人文、能源与环境、能源与未来、能源与变革、能源与气候变化等12个话题，通过审视与当今世界局势和社会生活密切相关的能源问题，引发学生对人类社会和福祉的深入思考。语篇多选自有影响力的国际学术期刊、学术著作和专业网站，以说明文、议论文为主，语言难度循序渐进。

针对大学生对提升英语综合应用能力和思辨能力的需求，本教材以能源与人类福祉相关内容为依托，精心设计讲解与活动，培养学生学术英语相关技能，提升思辨能力和自主学习能力，使学生逐渐过渡到能用英语学习相关领域的专业课程。

教材特色

选材兼具话题性与知识性。本教材从能源视角出发，找寻与当今社会热点话题的结合点，保证选材具有一定的知识性和讨论度，但又不涉及过于专业的能源技术知识，提高学生的学习积极性，保证良好的教学效果。

关注学术综合技能培养。本教材通过泛读、精读，指导学生掌握阅读学术文章的方法和技巧，培养学术阅读能力。在学术写作方面，从语句写作训练开始，培养学术写作基本功，强化正式语体意识；再到段落写作训练，培养段落布局、行文谋篇的能力，综合提升学术写作能力。同时，通过形式多样的活动设计，引导学生深入分析和评估所获取的信息，参与讨论并得体地评价他人观点，培养学生思辨能力，为从事学术活动打好基础。

提供立体化教学支持。本教材提供配套慕课资源，与纸质教材内容有机结合，实现线上线下立体化教学，为翻转课堂和混合式学习提供"抓手"。教师和学生可登录中国高校外语慕课平台 UMOOCs（moocs.unipus.cn），搜索"能源学术英语"，关注并学习慕课。本教材还配有电子版教师用书和助教课件，为教师备课和课堂教学提供有效支持。

教材结构

本教材分为两个级别，每个级别包含 6 个单元，每单元围绕一个能源热点问题展开，设置五大板块：视听说、泛读、精读、学术写作和单元任务。

Listening / Viewing and speaking：视听说板块。包含一段与单元主题相关的音频或视频素材，通过听力理解以及小型讨论活动，导入单元主题。

Extensive reading：泛读板块。包含一篇泛读文章，主题与单元话题相关。通过快速阅读的形式，帮助学生掌握文章大意，深入理解单元话题，为精读环节做好准备。

Intensive reading：精读板块。包含一篇精读文章，主题与泛读文章一致。通过把握文章结构和理解重点内容，训练学生的学术阅读能力。除此之外，从文章中精选重点学术词汇和表达，通过词汇含义、近义词、派生词、常用搭配、学术短语、句式结构、翻译及应用等训练，帮助学生全面、扎实地掌握学术语言特点。

Academic writing：学术写作板块。包含微观写作技能和宏观写作技能两个部分，分别关注句子和段落层面学术英语写作的特点，并配有丰富多样的练习。通过讲练结合的方式，提升学生的语言基本功和书面表达能力。

Sharing：单元任务板块。旨在综合运用单元所学内容完成产出任务。任务提供分步骤讲解，指导学生通过查阅资料、提炼信息、参与讨论等方式完成单元任务，并在课堂上以口头汇报形式与同学分享。

编写团队

《能源学术英语综合教程》编写团队来自中国石油大学（北京），主编是赵秀凤教授，负责教材整体策划和内容审定。

第一册全书汇编、修订由王文征完成。赵秀凤、王文征负责课文选材、题型设计。本册写作部分编写者为赵秀凤，听力部分编写者为蔡坤。各单元其余部分编写者为：第一单元王崴，第二单元李子男，第三单元连洁，第四单元王文征，第五单元李音，第六单元田芸、蔡坤。

第二册全书汇编、修订由陈芳完成。赵秀凤、王文征负责课文选材、题型设计。本册宏观写作技能部分编写者为徐方富，微观写作技能部分编写者为单小明，听力部分编写者为蔡坤，单元任务板块编写者为陈芳。各单元其余部分编写者为：第一单元李子男、王崴，第二单元陈芳、石虹，第三单元王文征，第四单元田芸、李音，第五单元郑世高，第六单元连洁。

中国石油大学（北京）外国语学院公共英语教学部其他任课教师也积极参与了教材编写，提出了宝贵意见，在此一并致谢。

编者
2023 年 6 月

Contents

Unit 1 Energy and Civilization	2
Viewing and speaking	3
Extensive reading	5
Reading text: Historical Per Capita Energy Consumption	5
Intensive reading	11
Reading text: The Culture of Energy: An Introduction	12
Academic writing	21
Sharing	27

Unit 2 Energy Access and Humanity	28
Viewing and speaking	29
Extensive reading	31
Reading text: Energy Access and Gender Equality	31
Intensive reading	37
Reading text: Energy Access and Health	37
Academic writing	46
Sharing	55

Unit 3 Energy and Environment	56
Listening and speaking	57
Extensive reading	59
Reading text: The Hidden Cost of Fossil Fuels	59
Intensive reading	67
Reading text: Global Warming: It's About Energy	67
Academic writing	77
Sharing	85

Unit 4   Energy and Future   86

Listening and speaking   87
Extensive reading   89
   Reading text: Gushing About America's Energy Future   89
Intensive reading   95
   Reading text: Install Residential Solar Panels and Save   96
Academic writing   106
Sharing   113

Unit 5   Energy and Transformation   114

Viewing and speaking   115
Extensive reading   117
   Reading text: Wind and Solar Power: Uncooperative Reality   117
Intensive reading   124
   Reading text: Is a US-style Shale Revolution Coming for Australia?   124
Academic writing   133
Sharing   139

Unit 6   Energy and Climate Change   140

Viewing and speaking   141
Extensive reading   143
   Reading text: What We Can Do to Reduce Climate Change   143
Intensive reading   150
   Reading text: Locking in the Long-term Vision   150
Academic writing   161
Sharing   169

Unit 1 Energy and Civilization

Lead-in

Energy is not only one of the major concerns of human beings nowadays, but a historically significant element in civilization. The history of human civilization, in one sense, can be seen as the history of energy development, from the primitive period to the industrial period, and finally to this highly technological era today. Do you know what major types of energy are consumed in each era? And how has people's life been shaped by energy?

In modern society, energy has become indispensable to people's life and has revolutionized their way of life. To understand energy's role in modern societies does not just involve the technological aspects. It requires an examination of the historical and cultural context. Can you name some of the cultural features of energy in the modern world?

Learning objectives

Upon completion of this unit, you will be able to:
- talk about the evolution of energy consumption in human history;
- express the role and features of energy in modern society;
- define key terms in writing;
- write an introductory paragraph to an essay;
- report the changes in energy use through three generations of your family.

Viewing and speaking

Viewing

New words

flirt /flɜːt/ v. 谈情说爱
biomass /ˈbaɪəʊˌmæs/ n. (用作燃料的) 生物质
nuclear /ˈnjuːklɪə/ a. 核能的
extract /ɪkˈstrækt/ v. 提取；取出
decompose /ˌdiːkəmˈpəʊz/ v. 腐烂；腐败
organism /ˈɔːɡəˌnɪz(ə)m/ n. 生物；有机体
release /rɪˈliːs/ v. 释放

fusion /ˈfjuːʒn/ n. 融合；聚变
depleted /dɪˈpliːtɪd/ a. 不足的；减少的
deterioration /dɪˌtɪərɪəˈreɪʃn/ n. 恶化
replenish /rɪˈplenɪʃ/ v. 补充；再充满
geothermal /ˌdʒiːəʊˈθɜːml/ a. 地热的
harness /ˈhɑːnɪs/ v. 利用
famine /ˈfæmɪn/ n. 饥荒
reliant /rɪˈlaɪənt/ a. 依赖的；依靠的

1 Watch the video clip and choose the best answer to each of the following questions.

1. Which activity is NOT mentioned to illustrate the importance of energy in daily lives?
   A) Working.   B) Traveling.   C) Eating.   D) Flying.
2. The disadvantages of fossil fuels do NOT include _____.
   A) depletion in decades   B) inconsistency and instability
   C) air quality deterioration   D) global warming
3. All the following are renewable energy resources EXCEPT _____.
   A) tides   B) wind   C) nuclear   D) sunlight

2 Watch the video clip again and fill in the blanks with what you hear.

1. Fossil fuels include _____, oil and _____.
2. As for the four major types of energy, fossil fuels consist of extracted _____ and plants of millions of years ago; biomass converts plants into _____ to produce energy; nuclear energy is released during _____; renewable energy comes from a source that's _____ when used, such as wind or solar power.
3. Rather than a one-time-use resource, resources for renewable energy can _____ in human lifetime. We don't need to compromise the earth to harness it, nor do we have to rely on other nations for these resources, which may lead to _____, _____ and political instability.
4. Presently the problem does not lie in technology, but in our _____ that is built around fossil fuel dependency. If we could convert it to _____, we will be able to rely on clean renewable energy.

Speaking

1 Read the short passage aloud and pay attention to the idea about primary energy and secondary energy.

Renewable and non-renewable energy sources are considered primary energy sources because they appear in the natural environment and can be used directly. Different resources of primary energy fall into two main categories: fuels and flows. Fuels are materials which contain one form of energy that can be transformed into another form of energy. Flows are natural processes that have energy associated with movement. Using a flow means harnessing energy that comes from this movement like wind and tides. Primary energy is energy that has not been obtained by anthropogenic conversion or transformation. The term "anthropogenic" is related to human activity or human influence.

Primary energy is often converted to secondary energy for more convenient use in human systems. Secondary energy sources are produced from primary sources of energy and can be used to store and deliver energy in a useful form. Hydrogen and electricity are considered secondary sources of energy, or "carriers" of energy.

2 Work in pairs and answer the two questions.

1. What are primary energy sources? How does primary energy get its name?
   _____
   _____

2. How would you define secondary energy based on the passage?
   _____
   _____

3 Do some research on energy in the Industrial period, a topic you will find in the texts in this unit, and then prepare a one-minute oral presentation. Make sure the messages are properly delivered to the audience.

Extensive reading

Reading text

Historical Per Capita Energy Consumption

1. The history of energy consumption shows how important energy is to the quality of life for each of us. Societies have depended on different types of energy in the past, and societies have been forced to change from one energy type to another. Global energy consumption can be put in perspective by considering the amount of energy consumed by individuals.

2. E. Cook (1971) provided estimates of daily human energy consumption at six different periods of societal development. Cook's estimates are given in Table 1.

Period	Era	Daily per capita consumption (1,000 kcal)				
		Food	H & C*	I & A**	Transportation	Total
Primitive	1 million BC	2				2
Hunting	100,000 BC	3	2			5
Primitive Agricultural	5000 BC	4	4	4		12
Advanced Agricultural	1400	6	12	7	1	26
Industrial	1875	7	32	24	14	77
Technological	1970	10	66	91	63	230

*H & C = Home and Commerce   **I & A = Industry and Agriculture

Table 1 Historical energy consumption (Cook, 1971)

The table shows that personal energy consumption has increased as society has evolved.

3. In ancient times, energy was consumed in the form of food. Cook assumed the only source of energy consumed by a person living during the period labeled "Primitive" was food. Energy is essential for life, and food was the first source of energy. According to Cook, humans require approximately 2,000 kilocalories

(about eight megajoules) of food per day. One food calorie is equal to one kilocalorie. One calorie is the amount of energy required to raise the temperature of one gram of water by one degree centigrade (°C). A change in temperature of 1°C is equal to a change in temperature of 1.8 degrees Fahrenheit (°F).

4   The ability to control fire during the Hunting period let people use wood to heat and cook. Fire provided light at night and could illuminate caves. Firewood was the first source of energy for consumption in a residential setting. Cook's estimate of the daily per capita energy consumption for Europeans about 100,000 years ago was 5,000 kilocalories (about 21 megajoules).

5   The Primitive Agricultural period was characterized by the domestication of animals. Humans were able to use animals to help them grow crops and cultivate their fields. The ability to grow more food than they needed became the impetus for creating an agricultural industry. Cook's estimate of the daily per capita energy consumption for people in the Fertile Crescent circa 5000 BC was 12,000 kilocalories (about 50 megajoules). Humans continue to use animals to perform work.

6   More energy was consumed during the Advanced Agricultural period when people learned to use coal, and built machines to harvest the wind and water. By the early Renaissance, people were using wind to push sailing ships, water to drive mills, and wood and coal for generating heat. Transportation became a significant component of energy consumption by humans. Cook's estimate of the daily per capita energy consumption for people in northwestern Europe circa 1400 was 26,000 kilocalories (about 109 megajoules).

7   The steam engine ushered in the Industrial period. It provided a means of transforming heat energy to mechanical energy. Wood was the first source of energy for generating steam in steam engines. Coal, a fossil fuel, eventually replaced wood and hay as the primary energy source in industrialized nations. Coal was easier to store and transport than wood and hay, which are bulky and awkward. Coal was useful as a fuel source for large vehicles, such as trains and ships, but of limited use for personal transportation. Oil, another fossil fuel, was a liquid and contained more energy per unit mass than coal. Oil could flow through pipelines and tanks. People just needed a machine to convert the energy in oil to a more useful form. Cook's estimate of the daily per capita energy consumption for people in England circa 1875 was 77,000 kilocalories (about 322 megajoules).

8   The modern Technological period is associated with the development of internal combustion engines, and applications of electricity. Internal combustion engines use oil and can vary widely in size. The internal combustion engine could be scaled to fit on a wagon and create "horseless carriages". The transportation

system in use today evolved as a result of the development of internal combustion engines. Electricity, by contrast, is generated from primary energy sources such as fossil fuels. Electricity generation and distribution systems made the widespread use of electric motors and electric lights possible. One advantage of electricity as an energy source is that it can be transported easily, but electricity is difficult to store. Cook's estimate of the daily per capita energy consumption for people in the United States circa 1970 was 230,000 kilocalories (about 962 megajoules).

Culture notes

megajoule: The joule (symbol: J) is a derived unit of energy in the International System of Units. The megajoule (MJ) is equal to one million joules. The energy required to heat 10 liters of liquid water at constant pressure from 0°C to 100°C is approximately 4.2 MJ. And one kilowatt hour of electricity is 3.6 MJ.

the Renaissance: It was the period in European history, especially Italy, from the 14th century to the 16th century. The period was conventionally characterized by a surge of interest in Classical scholarship and values. To the scholars and thinkers of the day, it was primarily a time of the revival of Classical learning and wisdom after a long period of cultural decline and stagnation.

Vocabulary

per capita /pə ˈkæpɪtə/ a. 人均的
advanced /ədˈvɑːnst/ a. 先进的；高级的
industrial /ɪnˈdʌstriəl/ a. 工业的；产业的
evolve /ɪˈvɒlv/ v. 逐步发展；逐渐演变
label /ˈleɪbl/ v. 贴标签于；把⋯⋯称为
kilocalorie /ˈkɪləˌkæləri/ n. 千卡
megajoule /ˈmegəˌdʒuːl/ n. 兆焦
calorie /ˈkæləri/ n. 卡路里
Fahrenheit /ˈfærənhaɪt/ n. 华氏温度
illuminate /ɪˈluːmɪˌneɪt/ v. 照亮；照明
characterize /ˈkærɪktəˌraɪz/ v. 描述⋯⋯的特征；描绘

domestication /dəˌmestɪˈkeɪʃn/ n. 驯养
cultivate /ˈkʌltɪˌveɪt/ v. 耕作；开垦
impetus /ˈɪmpɪtəs/ n. 推动力；促进因素
crescent /ˈkreznt/ n. 新月形
circa /ˈsɜːkə/ prep. 大约；左右
the Renaissance /rɪˈneɪsns; ˈrenəsɑːns/ n. 文艺复兴
sailing /ˈseɪlɪŋ/ n. 航行；航海
usher in 开启；开创
mechanical /mɪˈkænɪkl/ a. 机械的
hay /heɪ/ n. 干草
bulky /ˈbʌlki/ a. 体积大的；庞大的

pipeline /ˈpaɪpˌlaɪn/ n. 管道
internal /ɪnˈtɜːnl/ a. 内部的；里面的
combustion /kəmˈbʌstʃn/ n. 燃烧
internal combustion engine 内燃机

vary /ˈveəri/ v. 彼此相异
wagon /ˈwægən/ n. 四轮运货车
contrast /ˈkɒntrɑːst/ n. 差别；差异

▶ Integrated exercises

1 Match each of the following statements with the paragraph from which the information is derived. You may choose a paragraph more than once.

_____ 1 Coal became the major energy source in industrialized nations during the Industrial period.

_____ 2 According to Cook, food was the first and only source of energy during the Primitive period.

_____ 3 Energy matters a lot to people's quality of life, which can be perceived by reviewing the history of energy consumption.

_____ 4 People were able to use wood to heat and cook thanks to their ability of controlling fire.

_____ 5 The development of internal combustion engines contributes to today's transportation system.

_____ 6 More energy sources were found in wood and coal, wind and water.

_____ 7 E. Cook's table of daily personal energy consumption falls into six different periods of societal development.

_____ 8 The wide use of electric motors and electric lights was made possible by electricity generation and distribution systems.

_____ 9 People's ability to use animals to perform work characterized the Primitive Agricultural period.

_____ 10 Compared with coal, oil served as a better source of energy in terms of personal transportation.

2 Complete the following table to check your understanding of the major points and structure of the text.

Energy consumption in general (Paras. 1-2)	• The history of 1) _____ shows the importance of energy to our quality of life. Societies have been forced to change 2) _____ throughout the history owing to their dependence on energy. • A table of historical energy consumption made by E. Cook illustrates personal energy consumption has increased as society has 3) _____.
Energy consumption in detail (Paras. 3-8)	• In the Primitive period, 4) _____ was the only source of energy consumed by a person. • During the Hunting period, 5) _____ became the first source of energy in a residential setting thanks to people's ability to control fire. • In the Primitive Agricultural period, 6) _____ enabled humans to grow more food than they needed and this became the impetus for the creation of an agricultural industry. • During the Advanced Agricultural period, more energy was consumed because humans learned to use different forms of energy such as coal, wind and water. 7) _____ became a significant component of human energy consumption. • In the Industrial period, the 8) _____ was invented to transform heat energy to mechanical energy. Different types of energy were used in steam engines: wood and hay, coal and oil. • The modern Technological period is associated with the development of internal combustion engines that use 9) _____ and electricity is generated from 10) _____.

3 Work in pairs and list the major types of energy consumed in each historical period. Take notes of the energy types while you do this task.

_____
_____
_____

4 Translate the phrases into Chinese.

1 energy consumption _____
2 the quality of life _____
3 put ... in perspective _____
4 societal development _____
5 be essential for _____
6 be equal to _____
7 degree centigrade _____
8 degree Fahrenheit _____
9 source of energy _____
10 residential setting _____

11 be characterized by _____
12 domestication of animals _____
13 grow crops _____
14 cultivate fields _____
15 electric motor _____
16 electric light _____
17 perform work _____
18 sailing ship _____
19 generate heat _____

20 steam engine _____
21 mechanical energy _____
22 industrialized nation _____
23 unit mass _____
24 be associated with _____
25 internal combustion engine _____
26 primary energy _____

5 Complete the sentences by translating the Chinese given in brackets into English. Refer to the phrases listed above when necessary.

1 _____ (人均能源消耗) in the country is very low.
2 Understanding the pattern of positive and negative emotions will help you _____ (正确对待情感问题).
3 Water _____ (对于生物来说是必不可少的).
4 That may sound a lot, but it _____ (只相当于国民生产总值的 0.6%).
5 The city _____ (有着钢和玻璃结构的现代高层建筑).
6 It has been well documented that the cancer risks _____ (与吸烟有关).

Intensive reading

▶ Warming up

1 We talk a lot about energy consumption. How is energy consumption distributed globally? What is the major reason for the rapid growth of renewable energy consumption? The following passage may help clarify these questions.

As global energy consumption rises due to both the population and the demand per person for more energy, it presents two features.

The first feature is that the distribution of energy consumption globally is disproportionate. In 2017, China, the US and India were the largest consumers of primary energy globally. On a per capita basis, however, the data looks a bit different. As of 2015, Qatar, Bahrain and Kuwait were among the countries with the highest per capita energy consumption.

The other notable feature is that renewable energy consumption has increased dramatically in recent years and is projected to continue to increase. Renewable energy includes solar energy, wind power, hydroelectric energy, biomass and geothermal power, to name a few. By 2040, it is expected to reach about 2,748 million tonnes of oil equivalent. In comparison, the total renewable energy consumption totaled 35 million tonnes of oil equivalent in 1990.

Recent data suggests that renewable energy is growing as a share of the primary energy consumption worldwide. Among all countries globally, China has installed the most renewable energy capacity as of 2018. The growth of renewable energy has been largely due to reduction in technology costs. However, in order to reach a secure, sustainable and economically feasible energy system, global governments must implement policies to encourage and support renewable energy sources. Investment into renewables and a market design to reliably integrate renewables into modern infrastructure are necessary for successful implementation.

2 Work in groups and discuss how to reach a secure, sustainable and economically feasible energy system.

Reading text

The Culture of Energy: An Introduction

1. Energy is a hot issue these days. Primarily, this is because it is at the top of the list of concerns over the climate and the environment, the supply security, the price and economic growth. This is with good reason. Energy is an indispensable element in modern life, and if we look at the period after World War II, energy has increasingly become an essential part of everyday life. It has become one of the raw materials of human life, not the only one, but one of the most important.

2. Everyday life is unthinkable without energy. From the electric toothbrush's first oscillating motion to the bedside lamp's last ray of light, our life is based on energy use. Almost every activity we pursue during the day, on our way to work, at work, in the home or related to leisure – the activity presupposes energy use.

3. The culture of the modern world involves sizeable and continuous consumption of energy. Like most technology, energy use suspends some of the natural conditions of existence. The alteration between warmth and cold, between light and dark, has been suspended by the introduction of reliable heating and electric lighting in buildings. In most parts of the world, we have a light whenever we desire it, and the buildings maintain a comfortable temperature of 21 degrees Celsius (°C) by use of either heating or air conditioning. The welfare state has significantly sped up this development to a degree that notions such as wellness and individual well-being have become natural elements of our consumer culture.

4. The modern understanding of energy emerged with the industrial society. The First Industrial Revolution was indeed revolutionary exactly because mechanical energy replaced animal and human energy in the production of goods. Since the middle of the 19th century, modern energy found its way to other parts of society, to transport, the household, etc. In most – but not all – places, the gas light triumphed as the street light and the old natural oil lamps disappeared because of the brighter and safer light from the new lamps. As time passed, when the product was improved and the price became reasonable, the electric bulb was gradually the preferred choice. The city and the home became enlightened.

5. From the very beginning, the exploitation of gas and electricity was separate from the production of energy and from the use of energy in private households

and industry. Production and consumption of energy were slowly but surely separated and, in short, replaced by a structure in which production was located in the public sphere, while consumption remained in the private sphere. The introduction of district heating was somewhat later and a significant step in this process. After that, energy use turned into a question of pushing a button or turning a valve.

6 At the same time as energy has grown into a more indispensable element in modern life, to the individual human being, it has lost its materiality. This is a consequence of the centralization of production at power stations, at the combined heat and power stations, and at nuclear power plants. Subsequently, the word "energy" has become a symbol of the fact that we no longer have a direct connection to the value of fuels, but only to energy using instruments or appliances in manufacturing facilities, in households and on highways as well.

7 The disappearance of the present materiality of the fuels (wood, coke, coal, etc.) in favor of the absent materiality of energy, firmly indicates that a scientifically-based conception of the universe of energy requires more than an understanding of the technology-based improvement of everyday life and the technical and the economic sciences' quests for optimizing energy efficiency in production and consumption. We also need to address and explore the cultural contexts in which the daily turnings on and off of power and heat take place, if we want to reach a complex understanding of energy's role in modern societies. In other words, an understanding which not only focuses on the technological, political or economic aspects, but combines those perspectives with a thorough historical analysis of cultural contexts.

Culture notes

welfare state: It refers to the concept of government in which the state or a well-established network of social institutions plays a key role in the protection and promotion of the economic and social well-being of citizens. It is based on the principles of equality of opportunity, equitable distribution of wealth, and public responsibility for those unable to avail themselves of the minimal provisions for a good life.

the First Industrial Revolution: It is the process of change from an agrarian and handicraft economy to one dominated by industry and machine manufacturing. This process began in Britain in the 18th century and from there spread to other parts of the world.

Vocabulary

indispensable /ˌɪndɪˈspensəbl/ a. 必需的；不可或缺的
unthinkable /ʌnˈθɪŋkəbl/ a. 难以想象的
oscillate /ˈɒsɪˌleɪt/ v. 振荡；摆动
pursue /pəˈsjuː/ v. 从事；追求
leisure /ˈleʒə/ n. 消遣；娱乐
presuppose /ˌpriːsəˈpəʊz/ v. 以…为前提；预先假定
sizeable /ˈsaɪzəbl/ a. 相当大的
continuous /kənˈtɪnjʊəs/ a. 持续的；不间断的
suspend /səˈspend/ v. 暂停；中止；延缓
alteration /ˌɔːltəˈreɪʃn/ n. 变化；变动
Celsius /ˈselsiəs/ n. 摄氏温度
air conditioning n. 空调系统
significantly /sɪɡˈnɪfɪkəntli/ ad. 显著地
notion /ˈnəʊʃn/ n. 概念
wellness /ˈwelnəs/ n. 健康
well-being /ˌwelˈbiːɪŋ/ n. 健康；幸福

revolutionary /ˌrevəˈluːʃn(ə)ri/ a. 革命性的；突破性的
triumph /ˈtraɪʌmf/ v. 战胜；获胜
enlighten /ɪnˈlaɪtn/ v. 照耀；照亮
exploitation /ˌeksplɔɪˈteɪʃn/ n. 开发；利用
structure /ˈstrʌktʃə/ n. 结构；构造；体系
sphere /sfɪə/ n. 领域；范围
somewhat /ˈsʌmwɒt/ ad. 有点；稍微
valve /vælv/ n. 阀
materiality /məˌtɪərɪˈæləti/ n. 物质性；实体性
centralization /ˌsentrəlaɪˈzeɪʃn/ n. 集中
subsequently /ˈsʌbsɪkwəntli/ ad. 随后
disappearance /ˌdɪsəˈpɪərəns/ n. 消失
conception /kənˈsepʃn/ n. 概念；观念；想法
quest /kwest/ n. 追求；探索
context /ˈkɒntekst/ n. 环境；背景

▶ Text understanding

1 Read Para. 1 and answer the following questions.

1　Why is energy a hot issue today?

_____
_____

2　In what sense is energy an indispensable element in modern life?

_____
_____

3　What logic pattern does this paragraph employ?

_____
_____

2 Read Paras. 2-3 and find the examples illustrating the importance of energy use in modern life.

1 Energy use in daily life
_____
_____

2 Energy use in the natural conditions of existence
_____
_____

3 Complete the following table to check your understanding of the major points and structure of the text.

Introduction (Para. 1)	**The importance of energy:** • Energy has increasingly become a(n) 1) _____ part of everyday life. • Energy has become one of the 2) _____ of human life.
Body (Paras. 2-6)	**Energy use in modern society:** • Everyday life is 3) _____ without energy. • The culture of the modern world relies heavily on 4) _____.
	The historical progress of energy use: • The modern understanding of energy emerged with the 5) _____ when animal and human energy were replaced by 6) _____. • Modern energy found its way to other parts of society since 7) _____. • Production and consumption of energy were separated with production located in the 8) _____ and consumption in the 9) _____. • Energy has lost its 10) _____ as a result of the 11) _____ of production at power stations.
Conclusion (Para. 7)	Disappearance of materiality of energy calls for a complex understanding of the conception of energy, with a combined focus on the technological, political or economic aspects, along with 12) _____.

Energy and Civilization | 15

Language building

1 Match each of the words in the left column with its corresponding meaning in the right column.

_____	1 alteration	a	the full and effective use of sth.
_____	2 address	b	to obtain victory
_____	3 subsequently	c	to depend on sth. that is believed to exist or to be true
_____	4 exploitation	d	the act or process of changing sth.
_____	5 unthinkable	e	to give attention to or deal with a matter
_____	6 presuppose	f	impossible to imagine
_____	7 wellness	g	to move or swing from side to side regularly
_____	8 sizeable	h	the state of being healthy
_____	9 oscillate	i	fairly large
_____	10 triumph	j	after sth. else happened

2 Among the three choices given, choose the one that is **NOT** close in meaning to the underlined word in each sentence.

1. Energy is an <u>indispensable</u> element in modern life.
   A) necessary   B) essential   C) optional

2. This is a <u>consequence</u> of the centralization of production at power stations, at the combined heat and power stations, and at nuclear power plants.
   A) source   B) outcome   C) result

3. Like most technology, energy use <u>suspends</u> some of the natural conditions of existence.
   A) delays   B) stops   C) promotes

4. Almost every activity we <u>pursue</u> during the day presupposes energy use.
   A) engage in   B) guide   C) carry out

5. The buildings <u>maintain</u> a comfortable temperature of 21 degrees Celsius by use of either heating or air conditioning.
   A) keep   B) abandon   C) preserve

6. As time passed, when the product was improved and the price became reasonable, the electric bulb was gradually the <u>preferred</u> choice.
   A) favorite   B) preferable   C) ensured

7. The introduction of district heating was somewhat later and a <u>significant</u> step in this process.
   A) brilliant   B) important   C) great

8. Energy has <u>increasingly</u> become an essential part of everyday life.
   A) gradually   B) progressively   C) suddenly

9 Production and consumption of energy were replaced by a structure in which production was located in the public sphere, while consumption remained in the private sphere.

A) area   B) globe   C) field

10 In other words, an understanding which not only focuses on the technological, political or economic aspects, but combines those perspectives with a thorough historical analysis of cultural contexts.

A) concentrates on   B) looks upon   C) emphasizes

3 Complete each of the following sentences with an appropriate word from the given word family. Change the form where necessary.

1 The company makes a very _____ product. (rely, reliance, reliable)
2 The drug causes _____ of mood in some people. (oscillate, oscillation, oscillator)
3 Among the various social media policies adopted by businesses, one has risen to the top for its _____ view that social media use is an opportunity rather than a threat. (enlighten, enlightenment, enlightened)
4 On holidays, we can see a _____ line of cars on highways. (continue, continual, continuous, continuing)
5 Japanese industry is making _____ use of robots. (increase, increased, increasing, increasingly)
6 It is believed that education levels are strongly _____ to income. (relating, relation, related)
7 Her duties are _____ to keep the patients comfortable. (prime, primary, primarily)
8 If we cannot sell more goods, we'll have to cut back on the _____. (produce, product, production, producer)
9 These days, people are using a growing array of _____ communication methods – e-mail, instant messaging and social networking sites. (electric, electronic, electricity, electrified)
10 Some companies have suspended new oil drilling and _____ projects. (explore, exploration, explorer)

4 Match each word in the box with the group of phrases where it is usually found.

| maintain | issue | suspend | well-being | context | significant |

1 _____
cultural ~
historical ~
in the ~ of
put sth. in ~

2 _____
a sense of ~
mental ~
physical ~
economic ~

3 _____
~ the rescue work
~ negotiations with
~ talks
~ a project

4 _____
a hot ~
racial ~s
gender ~s
a key ~

5 _____
~ a comfortable temperature
~ a good relationship
~ order
~ contact with

6 _____
a(n) ~ step
~ development
a(n) ~ impact
a(n) ~ difference

5 Find the idiomatic expressions in the text matching the Chinese equivalents.

1 热门话题 _____
2 经济增长 _____
3 电动牙刷 _____
4 床头灯 _____
5 以…为基础 _____
6 明暗交替 _____
7 电力照明 _____
8 21 摄氏度 _____
9 自然生存条件 _____
10 加速 _____
11 个人幸福 _____
12 必不可少的元素 _____
13 发电站 _____
14 核电厂 _____
15 丧失物质性 _____
16 位于 _____
17 优化能效 _____
18 文化背景 _____
19 换言之 _____
20 关注 _____
21 与…有直接关系 _____
22 历史性分析 _____

6 Combine the sentences given below. Then compare your sentences with the original ones in the text.

1 _____

a In most parts of the world, we have a light whenever we desire it.
b The buildings maintain a comfortable temperature of 21 degrees Celsius by use of either heating or air conditioning.

2 _____

a The welfare state has significantly sped up this development to a degree.
b The degree is that notions such as wellness and individual well-being have become natural elements of our consumer culture.

3 _____

   a The word "energy" has become a symbol of a fact.
   b The fact is that we no longer have a direct connection to the value of fuels, but only to energy using instruments or appliances in manufacturing facilities, in households and on highways as well.

4 _____

   a We also need to address and explore the cultural contexts.
   b In the cultural contexts, the daily turnings on and off of power and heat take place, if we want to reach a complex understanding of energy's role in modern societies.

5 _____

   a In other words, an understanding which focuses on the technological, political or economic aspects.
   b The understanding combines those perspectives with a thorough historical analysis of cultural contexts as well.

7 Translate each of the Chinese sentences into English by using the underlined phrase or structure in the example.

1 Since the middle of the 19th century, modern energy <u>found its way</u> to other parts of society.
   他所发明的产品中只有一款进入市场销售。
   _____
   _____

2 From the electric toothbrush's first oscillating motion to the bedside lamp's last ray of light, our life <u>is based on</u> energy use.
   那部电视剧中的故事是基于现实生活的。
   _____
   _____

3 The disappearance of the present materiality of the fuels (wood, coke, coal, etc.) <u>in favor of</u> the absent materiality of energy, firmly indicates that a scientifically-based conception of the universe of energy requires more than an understanding of the technology-based improvement of everyday life.
   建造隧道的计划遭到否决，取而代之的是要架起一座桥梁。
   _____
   _____

4   In other words, an understanding which not only <u>focuses on</u> the technological, political or economic aspects, but combines those perspectives with a thorough historical analysis of cultural contexts.
这篇课文着重讲述能源发展的历史。
_____
_____

5   <u>From the very beginning</u>, the exploitation of gas and electricity was separate from the production of energy and from the use of energy in private households and industry.
孩子们从一开始就应该养成良好的习惯。
_____
_____

Academic writing

▶ Micro-skill: Definitions

Definitions are the explanation of an unfamiliar phrase, or a term used in a context, so as to make it clear to the reader.

1 Simple definitions

A simple definition is usually made up of the term being defined, the category it belongs to and its application.

Term	Category	Application
Electricity	is a basic part of nature	that is widely used as a form of energy.
A driver's license	is an official document	permitting a specific individual to operate one or more types of motorized vehicles, such as a motorcycle, car, truck or bus on a public road.
A photovoltaic module	is a packaged, connected assembly of solar panels	for the generation of electricity.

1 Complete the following definitions by choosing a suitable category word from the box.

vent   city   material   works   organs   instrument   behavior   molecule

1. A microscope is a(n) _____ used to see objects that are too small to be seen by the naked eye.
2. The lungs are the primary _____ of the respiratory system in humans and many other animals including a few fish and some snails.
3. A metropolis is a large _____, functioning as a significant economic, political and cultural center for a country or region, and an important hub for regional or international connections, commerce and communications.
4. Plastic is a(n) _____ consisting of any of a wide range of synthetic or semi-synthetic organic compounds that can be molded into solid objects.
5. Bullying is a pattern of anti-social _____ found in many schools.

6   A volcano is a(n) _____ in the crust of a planetary-mass object, such as the earth, that allows hot lava, volcanic ash and gases to escape from a magma chamber below the surface.

7   Literature refers to _____ or any single writing of artistic or intellectual value to be enjoyed by readers.

8   DNA is a(n) _____ used in the growth, development, functioning and reproduction of all known living organisms and many viruses.

2   Write definitions for the following terms.

1   Peking opera is _____
_____

2   Kidneys are _____
_____

3   Recycling is _____
_____

4   A metaphor is _____
_____

---

2   Complex definitions

A complex definition may explain or illustrate a term by
a) quoting a definition from another writer;
b) giving a variety of relevant situations;
c) explaining a process;
d) using a category word.

---

3   Read the following sentences and underline the term being defined. Then choose from a) to d) above to match each of the definitions.

_____   1   Fiction broadly refers to any narrative that is derived from the imagination – in other words, not based strictly on history or facts.

_____   2   An urban heat island (UHI) is an urban area or metropolitan area that is significantly warmer than its surrounding rural areas due to human activities with the temperature difference larger at night than during the day, especially when winds are weak.

_____   3   The Philippine poet Denn A. Meneses defines home as a magical place that "stays at the core of our being no matter where our life's journeys take us".

_____ 4 A community can be a city or town where people live with friends, neighbors, or in a church group, or a farm away from the distractions of the city, with like-minded people all willing to share their resources.

_____ 5 Are you afraid of high places or water? If so, you suffer from one of the most common phobias, which are strong, persistent, but unreasonable fears.

_____ 6 Acid rain is the result of tremendous quantities of fossil fuels such as coal and oil discharged annually into the atmosphere. Through a series of complex chemical reactions, these pollutants can be converted into acids, which may return to earth as components of either rain or snow.

3 Use of definitions

A definition can be used when a term in a title, even if it is quite common, needs to be defined to demonstrate the writer's understanding of its meaning. For example:

Should Universities Offer More Distance Learning Programs?
Distance learning programs are the online courses that students can attend to earn degrees and diplomas.

4 Read the following titles, underline the terms that may need defining and write definitions for them.

1 Is body language important to one's success in social contact?
_____
_____

2 Should the flow of immigrants be responsible for the crime and terrorism in Europe?
_____
_____

3 Will e-books replace printed books in the next 20 years?
_____
_____

Energy and Civilization | 23

▶ Macro-skill: Introductory paragraph

1 Overview of academic writing

Academic writing, by definition, is writing for academic purposes. While a little bit too "bookish" in the eyes of some people because they often assume that it only involves research, the word "academic" has a much broader meaning, just generally related to any academy or school, especially of higher learning. In other words, the learning activities of university students are mostly "academic", as opposed to "vocational" or "everyday".

As a kind of essay writing, which may involve narration, description, exposition and argumentation, academic writing focuses on the latter two genres and differs from vocational, everyday or literary writing in that it has its specific features in terms of essay organization / structure as well as lexical conventions and sentence patterns. For example, the first-person statement like "I think" should be avoided. A typical academic essay usually consists of three parts, i.e., introduction, main body and conclusion, though some may refer to the three parts as "beginning", "development" and "ending", which is more closely related to literary writing.

The introductory part often raises the question / issue / topic for discussion / development / analysis / debate to attract the attention of the reader. The main body of an essay develops / analyzes the issue or illustrates the idea raised in the introductory part in a certain logic, like problem-solution, classification, cause-effect or comparison and contrast. The concluding part of an essay sums up the analysis made in the main body by clarifying or stressing the view / comments / idea in a definite and forceful way to make a strong impression on the reader's mind.

2 How to write the introductory paragraph?

You can begin an essay in different ways to fit for the topic of discussion. Four ways of beginning an essay are given here for your reference.

1) Raising a question

2) Using the "but" pattern to introduce the writer's idea or topic for discussion

3) Presenting a fact / phenomenon or providing the background information

4) Quoting an idea or a saying

1 Read the following introductory paragraph and answer the question.

The English language is spoken or read by a large number of people in the world, for historical, political and economic reasons; but it may also be true that it owes something of its wide appeal to qualities and characteristics inherent in itself.

What are these characteristic features (which outstand in making the English language what it is), which give it its individuality and make it of this worldwide significance?

Question raised for discussion:

_____

_____

2 Read the following introductory paragraph and complete the table.

In 2016 the video gaming industry racked up sales of about $100 billion, making it one of the world's largest entertainment industries. The games range from time-wasting smartphone apps to immersive fantasy worlds in which players can get lost for days or weeks. Indeed, the engrossing nature of games is itself cause for concern. Last year four economists published a paper suggesting that high-quality video games – an example of what they call "leisure luxuries" – are contributing to a decline in work among young people, and especially young men. Given the social and economic importance of early adulthood, such a trend could spell big trouble. But are video games causing the young to drop out?

Topic introduced	In 2016 the video gaming industry became 1) _____. The games are 2) _____ and make players 3) _____. The engrossing nature of games is 4) _____. The idea of four economists is that 5) _____.
Topic for discussion	6) _____?

3 Read the following introductory paragraph and complete the table.

It seems likely that Teotihuacán's natural resources – along with the city elite's ability to recognize their potential – gave the city a competitive edge over its neighbors. The valley, like many other places in Mexican and Guatemalan highlands, was rich in obsidian (黑曜石). The hard volcanic stone was a resource that had been in great demand for many years, at least since the rise of the Olmecs (a people who flourished between 1200 BC and 400 BC), and it apparently had a secure market. Moreover, recent research on obsidian tools found at Olmecs sites has shown that some of the obsidian obtained by the Olmecs originated near Teotihuacán. Teotihuacán obsidian must have been recognized as a valuable commodity for many centuries before the great city arose.

Topic for discussion	Teotihuacán's competitive edge
Topic to be developed	Teotihuacán's natural resources gave the city 1) _____.
Facts provided	The valley was 2) _____. Obsidian 3) _____ for many years, and 4) _____ for many centuries.

4 Read the following introductory paragraph and complete the table.

Hegel, the German philosopher, says, "We learn from history that men never learn anything from history." This wry remark has been confirmed time and again by historical events, one of which is Hitler's invasion of the Soviet Union. He must have clean forgotten or willfully ignored the great disaster Napoleon brought upon himself by attacking Russia in the 19th century.

Topic for discussion	Hitler's invasion of the Soviet Union
Quotation	1) _____ _____.
Idea to convey	Hitler's failure in his invasion of the Soviet Union is similar to 2) _____ _____.

--- **Writing assignment** ---

Write an introductory paragraph on the following topics. Each paragraph should be around 50 words in an academic or written style.

1 Energy is indispensable in modern society.
2 Energy is the engine of civilization.

Sharing

Tell the story of how your family uses energy and report the changes in energy use through three generations of your family.

1 Conduct an interview with your parents and grandparents. Ask them the following questions:

1) What types of energy did your family use when you were young?
2) How did you use them?
3) What was your life like at that time?

2 Work in groups and summarize the information from your interview and prepare a group oral presentation with PPT including the three parts:

1) "My family" (basic information about your family)
2) "Energy story of my family" (energy types used by and the life conditions of your grandparents, your parents and yourself)
3) Conclusion (how energy has changed people's life)

3 Give your presentation in class.

Unit 2 Energy Access and Humanity

Lead-in

Energy has a great impact on humanity issues, such as gender equality and human health. People may not be aware of the fact that a lack of modern energy access harms women more than men, and to achieve gender equality, one of the most effective ways is to ensure reliable sources of energy. Do you know why?

One of the ways in which women suffer from poor energy access is the household air pollution related to cooking and other household chores. There are also other impediments imposed on people's health by energy poverty with no gender difference, affecting males and children, too. What are these impediments?

Learning objectives

Upon completion of this unit, you will be able to:
- explain the relationship between energy access and gender equality;
- illustrate the impact of energy access on health;
- use examples properly and effectively in academic writing;
- recognize and use the pattern of problem-solution in writing;
- make a proposal on how your local community improves gender equality and health by upgrading energy access.

Viewing and speaking

▶ Viewing

New words

Senegal /ˌsenɪˈɡɔːl/ 塞内加尔
Dakar /dəˈkɑː/ 达喀尔（塞内加尔首都）
charcoal /ˈtʃɑːˌkəʊl/ n. 木炭
franc /fræŋk/ n. 西非法郎（FCFA）
entrepreneur /ˌɒntrəprəˈnɜː/ n. 企业家
strip /strɪp/ v. 剥去；除去
Senegalese /ˌsenɪɡəˈliːz;ˌsenɪɡəˈliːs/
  n. 塞内加尔人
  a. 塞内加尔的
in tandem 协同工作
livelihood /ˈlaɪvlihʊd/ n. 生计

1 Watch the video clip and put a tick before the changes that the energy programs have brought about in Senegal.

☐ 1 new cook stoves
☐ 2 improved charcoal production
☐ 3 education for children
☐ 4 international trade
☐ 5 a solar-powered mill
☐ 6 efficient transportation

2 Watch the video clip again and fill in the blanks with what you hear.

1 Women and their dependent children make up the _____ of those living in poverty. So energy programs here are now being designed to include women in productive activities and _____ roles.

2 But far from Dakar, the project is also helping improve lives of _____ women and helping protect the environment. While consumers are using less charcoal, _____ are making more money.

3 With recent charcoal sales, Marietou bought _____ and now makes extra money for her family by selling frozen treats. With electricity, she and other women have _____ opportunities to generate additional income.

4 Remote areas are being connected to the national grid, allowing women and their families greater _____ to a number of services and to a wider world. The country as a whole is relying more on sources of energy that are clean and _____.

5 From a new cook stove to improved charcoal production, to better livelihoods and stronger economy, Senegal is _____ its economy, its _____ and the lives of its people, especially of women who are major consumers of energy and are key to pulling families up from poverty.

Speaking

1 Read the short passage aloud and pay attention to the idea about energy conversation.

My daughter and I had a conversation about energy. I talk with Ally about energy because no one did this with me when I was her age. In the 1980s, we just got into our cars and took off. We took trips on airplanes and we went along our merry way and consumed energy. No one even gave it a second thought, unless of course we went to war or there was a chemical spill. The news media didn't care unless there was something sensational to report.

Fast forward to 2017, it wasn't until recently that the energy conversation really started so openly. I'd like to think social media helped us get to this place. With women over 50% of the workforce, we need to have these conversations and encourage more education and awareness about the opportunities the industry provides to communities and the workforce at large.

2 Work in pairs and answer the two questions.

1. According to the narrator, what was people's attitude toward energy consumption in the 1980s?

   _____
   _____

2. What need people do about energy when it comes to this day? Why?

   _____
   _____

3 Do some research on the relationship between energy access and gender equality, a topic you will find in the texts in this unit, and then prepare a one-minute oral presentation. Make sure the messages are properly delivered to the audience.

Extensive reading

Reading text

Energy Access and Gender Equality

1   There is a clear and important intersection between access to energy and gender equality. While a lack of access to modern and sustainable forms of energy impacts both men and women, inequalities in social standing, economic capability and gender-defined roles mean women often suffer from a lack of energy access in a disproportionate way. In developing countries, women tend to bear responsibility for collecting and preparing fuel for cooking, as well as the cooking itself. Households dedicate an average of 1.4 hours a day collecting fuel, a burden borne mainly by women and children (Figure 1). Moreover, according to UNEP, the loads that they carry can have an impact on their physical well-being; in Africa, women carry loads that weigh as much as 25-50 kilograms. A lack of access to clean, modern cook stoves and fuels, and a reliance on the traditional use of biomass for cooking also mean that women (and children) are the ones most impacted by household air pollution. In rural households in South Asia, it is estimated that women spend four hours a day cooking when using traditional stoves. Women also bear most of the responsibility for household tasks such as cleaning and clothes washing, which take longer to do without access to modern energy.

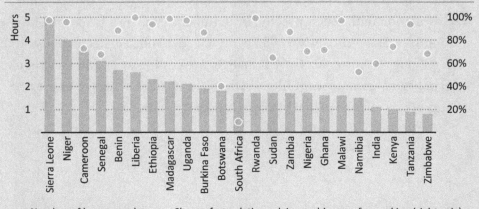

Figure 1 Average number of hours spent collecting fuel per day per household

Energy Access and Humanity | 31

2. Providing access to modern, affordable and reliable sources of energy has the potential to improve the well-being of women and children, and to provide women with new economic opportunities. Time saved through access to modern energy can be redirected toward education, social and family activities, and economic opportunities. Education has been shown to be an important driver of poverty reduction and income generation, and to enhance gender equality. Over 90 million primary school aged children in sub-Saharan Africa attend schools without electricity. In Brazil, girls with access to electricity in rural areas are almost 60% more likely to finish primary school by the age of 18 than those without, and self-employed women with access to energy have incomes that are two times higher than those without. In South Africa, electrification has raised female employment in newly electrified communities by almost 10% because it has improved the efficiency of achieving household tasks. In Nicaragua, access to reliable electricity has increased the ability of rural women to work outside the home by 23%.

3. Women are often best positioned to identify, champion and deliver energy access solutions. The evidence suggests there are significant advantages in involving women from start to finish in the design of modern energy access technologies and programs, and empowering women to become more involved in the provision of energy services. This is in part because they hold specific local knowledge about how different options could serve household needs, and because of their ability to influence their peers.

4. Despite this, there is further to go in routinely including women when designing and implementing energy access strategies and ensuring that policies take gender into account. A number of multinational development banks, regional organizations and bilateral donors are now taking steps to mainstream gender in energy programs and policy and methodologies and best practices are beginning to emerge. Moreover, several initiatives now bring together energy, gender, health and climate with women's empowerment, employment and representation in the energy sector, including the Global Alliance for Clean Cookstoves, SEforALL, ECOWAS Energy-Gender Policy & Regulation and the Clean Energy Ministerial. These initiatives are positive steps and they should help policymakers and institutions better identify key barriers to improving energy access and the design of effective access programs.

Culture notes

UNEP (the United Nations Environment Programme): It is the leading global environmental authority that sets the global environmental agenda, promotes the coherent implementation of the environmental dimension of sustainable development within the United Nations system, and serves as an authoritative advocate for the global environment.

Global Alliance for Clean Cookstoves: It is a public-private partnership that seeks to save lives, improve livelihoods, empower women and protect the environment by creating a thriving global market for clean and efficient household cooking solutions.

Vocabulary

gender /ˈdʒendə/ n. 性别
intersection /ˈɪntəˌsekʃn/ n. 交叉；交集
inequality /ˌɪnɪˈkwɒləti/ n. 不平等；不公平
disproportionate /ˌdɪsprəˈpɔːʃnət/ a. 不均衡的；不成比例的
kilogram /ˈkɪləˌɡræm/ n. 公斤；千克
self-employed /ˌself ɪmˈplɔɪd/ a. 个体经营的；自由职业的
electrification /ɪˌlektrɪfɪˈkeɪʃn/ n. 电气化
employment /ɪmˈplɔɪmənt/ n. 雇用；就业；受雇
electrified /ɪˈlektrɪˌfaɪd/ a. 通电的

peer /pɪə/ n. 同龄人；社会地位相同的人
routinely /ruːˈtiːnli/ ad. 惯常地；例行地
multinational /ˌmʌltiˈnæʃn(ə)l/ a. 跨国的；多国的
bilateral /baɪˈlæt(ə)rəl/ a. 双边的
methodology /ˌmeθəˈdɒlədʒi/ n. 方法；方法论
empowerment /ɪmˈpaʊəmənt/ n. 赋权；自主
representation /ˌreprɪzenˈteɪʃn/ n. 代表
ministerial /ˌmɪnɪˈstɪəriəl/ a. 部长的

▶ Integrated exercises

1 Match each of the following statements with the paragraph from which the information is derived. You may choose a paragraph more than once.

_____ 1 Women, in developing countries, tend to be responsible for cooking and collecting and preparing fuel for cooking.

_____ 2 The income of self-employed women with access to energy in Brazil is three times that of those without.

_____ 3 According to the evidence, we will have significant advantages if we engage women in the design of modern energy access technologies and programs from beginning to end.

_____ 4 Women are likely to be the ones most affected by household air pollution due to a lack of access to clean, modern cook stoves and fuels and a reliance on the traditional use of biomass for cooking.

_____ 5 Time saved through the use of modern energy can be reused for education, social and family activities, and economic opportunities.

_____ 6 Women have to spend more time doing household work such as cleaning and clothes washing without access to modern energy.

_____ 7 There are now several initiatives that combine energy, gender, health and climate with women's empowerment, employment and representation in the energy sector.

_____ 8 Partially because women specifically know about how different options meet household needs, and because they are able to influence their peers, women are often the most suitable for identifying, championing and delivering energy access solutions.

_____ 9 Many multinational development banks, regional organizations and bilateral donors are taking measures to mainstream gender in energy programs.

_____ 10 The well-being of women and children can be potentially improved if they have better access to modern, affordable and reliable sources of energy.

2 Complete the following table to check your understanding of the major points and structure of the text.

Introduction to the problem (Para. 1)	There is a clear and important intersection between access to 1) _____ and 2) _____. Inequalities in social standing, economic capability and gender-defined roles mean women often suffer from 3) _____ in a disproportionate way.
Benefits of problem solving (Para. 2)	Providing access to modern, affordable and reliable sources of energy has the potential to improve the 4) _____, and to provide women with new economic opportunities.
Solutions and conclusions (Paras. 3-4)	• Women are often best positioned to identify, champion and deliver 5) _____. This is in part because they hold specific local knowledge about 6) _____, and because of 7) _____. • There is further to go in routinely including women when designing and implementing 8) _____ and ensuring that policies take 9) _____ into account. Several initiatives now bring together energy, gender, health and climate with 10) _____ in the energy sector.

3 Complete the following table with the evidence of impacts of a lack of access to modern energy and benefits of providing access to modern energy.

Impacts of a lack of access to modern energy	1) _____ 2) _____ 3) _____ 4) _____
Benefits of providing access to modern energy	1) _____ 2) _____ 3) _____ 4) _____

4 Work in pairs and give examples to demonstrate the impacts of energy poverty on women. Take notes of the examples while you do this task.

_____
_____
_____

Energy Access and Humanity

5 Translate the phrases into Chinese.

1. access to energy _____
2. it is estimated that _____
3. economic capability _____
4. a lack of energy access _____
5. bear responsibility for _____
6. physical well-being _____
7. household air pollution _____
8. economic opportunity _____
9. multinational development bank _____
10. income generation _____
11. household task _____
12. in part _____
13. bilateral donor _____
14. gender equality _____
15. social standing _____
16. gender-defined role _____
17. in a disproportionate way _____
18. modern cook stove _____
19. be best positioned to _____
20. poverty reduction _____
21. newly electrified community _____
22. reliable electricity _____
23. regional organization _____
24. energy sector _____

6 Complete the sentences by translating the Chinese given in brackets into English. Refer to the phrases listed above when necessary.

1. Women _____ (往往要承担照顾家庭的责任).
2. Dieting often _____ (对女性的身体健康有影响).
3. _____ (据估计) over one million species are on the verge of extinction.
4. A significant portion of the world's population _____ (苦于缺乏能源).
5. Implementation of specific economic and social policies has been shown to be _____ (减贫和创收的重要推动力).
6. They are _____ (处于有利位置) help ease the water-collection burden on women in this area.

Intensive reading

▶ Warming up

1 Complete the description of the picture below by following the clues provided.

1 Cooking Fire = 1 Automobile

This picture illustrates 1) _____. In the upper part of the picture, there is a(n) 2) _____ which releases misty smoke. What can be seen in the lower part are 3) _____ and 4) _____, with an equation saying "one cooking fire equals one automobile", which conveys the shocking information that 5) _____.

2 Based on Task 1, make a brief comment on the relationship between energy and health.

▶ Reading text

Energy Access and Health

1   Access to energy services is critical for advancing human development, furthering social inclusion of the poorest and most vulnerable in society and meeting many of the SDGs. In September 2015, 193 countries – developing and developed countries alike – adopted the Sustainable Development Goals, known officially as the 2030 Agenda for Sustainable Development. The 17 new SDGs aim

at ending poverty, improving health and gender equality, protecting the planet, and ensuring peace and prosperity for all. For the first time, the SDGs include a target focused specifically on ensuring access to affordable, reliable and modern energy for all by 2030 (SDG 7.1), signaling a recognition of the importance of access to modern energy services in its own right, and of the centrality of energy in achieving many of the other development goals. The SDGs recognize the integrated nature of development. A lack of access to modern energy can make it difficult or impossible for a country to confront the myriad challenges that it faces, such as poverty (SDG 1), air pollution, low levels of life expectancy and lack of access to essential healthcare services (SDG 3), delivering quality education (SDG 4), adaptation and mitigation of climate change (SDG 13), food production and security (SDG 2), economic growth and employment (SDG 8), sustainable industrialization (SDG 9) and gender inequality (SDG 5).

2   Health is one of these areas where there are important linkages with energy access. The use of candles, kerosene and other polluting fuels for lighting has serious implications for health, so does reliance on solid biomass and coal for cooking (often in an enclosed space without proper ventilation). Together, they are responsible for an estimated 2.8 million premature deaths per year worldwide.

3   Efforts to improve access to clean cooking have immense potential to improve household air quality and health, especially for women and children. The incomplete combustion of solid biomass in a three-stone fire, which is the most common traditional cooking method, releases significant particulate matter. There are alternatives such as improved or advanced biomass cook stoves (which have a chimney or a fan to aid combustion), and stoves fueled by LPG, natural gas or solar power, which reduce household air pollution. Yet, even when these are available, households may decide they cannot afford them; or that the increased costs are not worthwhile, even if they are affordable. Moreover, the least-cost "improved" solution is not emissions free, and often used alongside traditional alternatives, rather than entirely superseding them (known as fuel stacking).

4   While the challenges of providing access to clean cooking are many, so are the benefits. Clean cooking provides direct health benefits, including a reduction in the number of premature deaths. It also has the potential to help deliver other SDGs by reducing greenhouse gas emissions and improving the lives of women and children through a reduction in the burden of household chores related to fuel collection and cooking. In short, there are significant synergies between policies to address energy access, local air pollution, health and climate change, which underline the importance of integrating policies and local initiatives to reduce barriers to improving access to clean cooking.

5   There are other synergies between energy access and health. Many people take for granted that healthcare facilities depend on energy to function and provide essential services. Almost 60% of health facilities in sub-Saharan Africa have no electricity while on average, just 34% of hospitals and 28% of health facilities in sub-Saharan Africa have reliable electricity access. An estimated 60% of refrigerators used in health clinics in Africa have unreliable electricity, resulting in a loss of almost half of vaccines, while 70% of electrical medical devices used in developing countries fail, with poor power quality a major contributing factor. Additionally, a lack of electricity means medical staff often has to work with flashlights or kerosene lamps. Moreover, the energy needs of the health sectors in low- and middle-income countries are expected to rise. The need for cold storage for vaccines, for example, is expected to increase and the growing need to fight diseases requires complex interventions that will drive additional energy requirements.

Culture notes

Sustainable Development Goals (SDGs): They were adopted by all United Nations Member States in 2015 as a universal call to action to end poverty, protect the planet and ensure that all people enjoy peace and prosperity by 2030. They address the global challenges we face, including those related to poverty, inequality, climate, environmental degradation, prosperity, and peace and justice.

LPG: Liquefied petroleum gas or liquid petroleum gas (LPG or LP gas) is a flammable mixture of hydrocarbon gases used as fuel in heating appliances, cooking equipment and vehicles.

Vocabulary

inclusion /ɪnˈkluːʒn/ n. 包含；包括
prosperity /prɒˈsperəti/ n. 繁荣；昌盛
specifically /spəˈsɪfɪkli/ ad. 专门地
recognition /ˌrekəɡˈnɪʃn/ n. 认可；承认
centrality /senˈtræləti/ n. 中心地位
integrated /ˈɪntɪˌɡreɪtɪd/ a. 综合的
confront /kənˈfrʌnt/ v. 面对；正视
myriad /ˈmɪriəd/ a. 无数的；极大量的
life expectancy n. 预期寿命
healthcare /ˈhelθˌkeə/ n. 医疗保健
industrialization /ɪnˌdʌstriəlaɪˈzeɪʃn/ n. 工业化
linkage /ˈlɪŋkɪdʒ/ n. 联系；关联
enclosed /ɪnˈkləʊzd/ a. 封闭的

ventilation /ˌventɪˈleɪʃn/ n. 通风；通风系统
responsible /rɪˈspɒnsəbl/ a. 负有责任的
premature /ˈpremətʃə/ a. 过早的；提早的
immense /ɪˈmens/ a. 巨大的
incomplete /ˌɪnkəmˈpliːt/ a. 不完全的
particulate /pɑːˈtɪkjʊleɪt/ a. 微粒的
chimney /ˈtʃɪmni/ n. 烟囱；烟道
supersede /ˌsuːpəˈsiːd/ v. 取代；替代
stack /stæk/ v. 堆积；堆放
chore /tʃɔː/ n. 家庭杂务
synergy /ˈsɪnədʒi/ n. 协同；协同作用
unreliable /ˌʌnrɪˈlaɪəbl/ a. 不可靠的；易出问题的
additionally /əˈdɪʃn(ə)li/ ad. 此外；另外

▶ Text understanding

1 Read Para. 1 and complete the summary of the paragraph.

What is the significance of the 2030 Agenda for Sustainable Development?
It is the 1) _____ time for the SDGs to include a(n) 2) _____ specifically focused on ensuring access to affordable, reliable and modern energy for all by 2030, and to recognize how 3) _____ the access to modern energy services is in its own right, and how 4) _____ energy is in achieving many other 5) _____.

2 Read Paras. 2-5 and complete the following table.

Impact of poor energy access on health (Para. 2)	Among other areas, health has important 3) _____ energy access. • Polluting energy sources for 4) _____ and 5) _____ have serious implications for health, leading to a large number of 6) _____ per year worldwide.
1) _____ in problem solving (Para. 3)	• New alternatives may not be 7) _____ or the increased costs are not 8) _____. • The least-cost "improved" solution is not 9) _____, and often used 10) _____.

(to be continued)

(continued)	
2) _____ from problem solving (Paras. 4-5)	11) _____ between energy access and health: • Clean cooking provides direct health benefits, including a reduction in 12) _____. • Clean cooking reduces 13) _____. • Clean cooking reduces the burden of 14) _____. • 15) _____ depend on energy to function and provide essential services.

▶ Language building

1 Match each of the words in the left column with its corresponding meaning in the right column.

_____	1 vulnerable	a	a new action for dealing with a particular problem
_____	2 myriad	b	a pipe through which smoke goes up into the air
_____	3 biomass	c	weak and easily hurt physically or emotionally
_____	4 premature	d	a substance put into the body to protect against a disease
_____	5 combustion	e	the process of burning
_____	6 particulate	f	a large number of sth.
_____	7 chimney	g	to show sth. is important
_____	8 initiative	h	plant and animal waste used as fuel
_____	9 underline	i	happening before the natural or appropriate time
_____	10 vaccine	j	relating to or consisting of very small pieces of a substance

2 Among the three choices given, choose the one that is **NOT** close in meaning to the underlined word in each sentence.

1 Access to energy services is <u>critical</u> for advancing human development.
   A) crucial   B) secure   C) key

2 The SDGs include a target focused specifically on ensuring access to affordable, reliable and modern energy for all by 2030 (SDG 7.1), <u>signaling</u> a recognition of the importance of access to modern energy services in its own right.
   A) indicating   B) signing   C) suggesting

3 The incomplete combustion of solid biomass in a three-stone fire, which is the most common traditional cooking method, <u>releases</u> significant particulate matter.
   A) sets free   B) gives out   C) relieves

4 There are alternatives such as improved or advanced biomass cook stoves (which have a chimney or a fan to <u>aid</u> combustion).
   A) hamper   B) help   C) assist

5 Yet, even when these are <u>available</u>, households may decide they cannot afford them.
   A) obtainable   B) at hand   C) avoidable

6 Moreover, the least-cost "improved" solution is not emissions free, and often used alongside traditional alternatives, rather than entirely superseding them.
A) replacing B) outpacing C) substituting for

7 A lack of access to modern energy can make it difficult or impossible for a country to confront the myriad challenges that it faces.
A) face B) conform C) handle

8 In short, there are significant synergies, which underline the importance of integrating policies and local initiatives to reduce barriers to improving access to clean cooking.
A) obstacles B) hurdles C) hazards

9 The growing need to fight diseases requires complex interventions that will drive additional energy requirements.
A) impel B) stimulate C) ride

10 Clean cooking provides direct health benefits, including a reduction in the number of premature deaths.
A) redundancy B) decrease C) decline

3 Complete each of the following sentences with an appropriate word from the given word family. Change the form where necessary.

1 Our aim is to create a fairer, more _____ society. (include, including, inclusive, inclusion)

2 Blacks have been hurt by racial _____ in housing and education. (equal, equity, equality, inequality)

3 There are some areas of poverty, but the country as a whole is fairly _____. (prosper, prosperous, prosperity)

4 The _____ of the German economy to its welfare system must be recognized. (center, centered, central, centrality)

5 This computer program can be _____ with existing programs. (integrating, integrated, integrative, integration)

6 Some novels quite readily lend themselves to _____ as plays. (adapt, adapted, adaptive, adaptation, adapter)

7 Reading between the lines means catching the _____ meaning in the text. (imply, implied, implication)

8 The _____ of the universe is difficult to grasp. (immense, immensely, immensity)

9 Lesley had agreed to Jim going skiing in _____ years. (alternate, alternative, alternation)

10 He was the _____ and pioneer of modern Chinese prose. (initiate, initiative, initiation, initiator)

4 Match each word in the box with the group of phrases where it is usually found.

| function | deliver | additional | adopt | barrier | release |

1 _____
 ~ a policy
 ~ an approach
 ~ a new method
 ~ a suggestion

2 _____
 ~ energy requirements
 ~ support
 ~ income
 ~ cost

3 _____
 ~ quality education
 ~ improved services
 ~ a lecture
 ~ an attack

4 _____
 ~ particulate matter
 ~ emissions
 ~ heat
 ~ toxic gas

5 _____
 ~ normally
 ~ properly
 ~ efficiently
 ~ as sth.

6 _____
 ~ to sth.
 language ~
 remove ~s
 break down ~s

5 Find the idiomatic expressions in the text matching the Chinese equivalents.

1 推动人类发展 _____
2 促进社会包容性 _____
3 消除贫穷 _____
4 促进性别平等 _____
5 保障和平与繁荣 _____
6 发展目标 _____
7 基本医疗服务 _____
8 提供优质教育 _____
9 气候变化减缓 _____
10 可持续工业化 _____
11 封闭空间 _____
12 过早死亡 _____
13 巨大潜力 _____
14 微粒物质 _____
15 无排放 _____
16 强调…的重要性 _____
17 当地举措 _____
18 主要成因 _____
19 医护人员 _____
20 中低收入国家 _____
21 冷藏 _____
22 抗击疾病 _____

6 Combine the sentences given below. Then compare your sentences with the original ones in the text.

1 _____
_____

 a In September 2015, 193 countries – developing and developed countries alike – adopted the Sustainable Development Goals.
 b The Sustainable Development Goals are known officially as the 2030 Agenda for Sustainable Development.

2 _____

    a The use of candles, kerosene and other polluting fuels for lighting has serious implications for health.

    b Reliance on solid biomass and coal for cooking (often in an enclosed space without proper ventilation) also has serious implications for health.

3 _____

    a The incomplete combustion of solid biomass in a three-stone fire releases significant particulate matter.

    b The incomplete combustion of solid biomass in a three-stone fire is the most common traditional cooking method.

4 _____

    a In short, there are significant synergies between policies to address energy access, local air pollution, health and climate change.

    b These significant synergies underline the importance of integrating policies and local initiatives to reduce barriers to improving access to clean cooking.

5 _____

    a Healthcare facilities depend on energy to function and provide essential services.

    b Many people take this for granted.

7 Translate each of the Chinese sentences into English by using the underlined phrase or structure in the example.

1 … signaling a recognition of the importance of access to modern energy services <u>in its own right</u> …
虽然这部电影是根据一部畅销小说改编的，但它本身就很棒。

_____

2 Health is one of these areas where there are <u>important linkages with</u> energy access.
本章展示了经济发展与环境之间有重要联系。

_____

3 The use of candles, kerosene and other polluting fuels for lighting <u>has serious implications for</u> health.
它对于经济发展有严重的影响。

_____

4 In short, there are significant synergies between policies to address energy access, local air pollution, health and climate change.
这两家公司间显然有重大的协同效应。
_____
_____

5 On average, just 34% of hospitals and 28% of health facilities in sub-Saharan Africa have reliable electricity access.
平均来看，该地区的排放量每立方米超出了 100 微克。
_____
_____

6 An estimated 60% of refrigerators used in health clinics in Africa have unreliable electricity resulting in a loss of almost half of vaccines.
不遵守这些条件将导致合同终止。
_____
_____

7 The need for cold storage for vaccines, for example, is expected to increase.
预计这项法案将在下一届议会中通过。
_____
_____

Academic writing

▶ Micro-skill: Examples

Examples are used in academic writing to strengthen the argument or help the reader understand a point.

1 The use of examples

When an idea needs to be illustrated, examples should be used.

People need to eat many different foods to stay healthy. Vegetables, for example, can provide various vitamins to human bodies.

2 Expressions to introduce examples

1) for instance, for example (with commas)

   Fossil fuels, for instance coal, can cause serious pollution.

2) such as

   Many great scientists such as Newton loved to raise all kinds of questions when they were young.

3) particularly, especially (to give a focus)

   Certain people, especially young people, are inclined to have big dreams.

4) a case in point (for single examples)

   It's important to always encrypt your data so hackers can't steal it. The recent data breach is a case in point.

1 Read the example and add a suitable example to each sentence. Introduce it with one of the expressions above.

Example:
   Knowing a foreign language is very helpful in one's career.
   → Knowing a foreign language, for instance English, is very helpful in one's career.

1 The invention of modern transportation affected the lives of most people.

2 Since the late 1990s many countries have introduced fees for university courses.

3 Girls in many countries have more opportunities to obtain education than before.

4   The doctor examined various parts of my body before diagnosing my illness as bronchitis.

5   Meeting the needs of future generations may need action now.

2 Read the following passage and add examples from the box where suitable, using the expressions denoting examples.

| many schools offer online courses | flipped classrooms |
| free study time   classrooms | teachers and tutors |

With the widespread use of the Internet, the ways of study have undergone some significant changes. It is no longer necessary to go to classrooms to study a regular course. With specially designed courses, students can study at home or any other places away from their schools. Such online courses can provide a lot of convenience to students. As a result, certain facilities in schools are not in great demand. However, other components of a regular school are still needed for supervision and guidance, so it seems unlikely that the Internet will completely replace regular schools.

3 Complete the following passage by adding appropriate supporting examples.

Computers have become an essential part of modern human life. Since the invention of computers, they have evolved in terms of increased computing power and decreased size. Owing to the widespread use of computers in every sphere, life in today's world would be unimaginable without computers. They have made human life better and happier. There are many uses of computers in different fields of work.

Macro-skill: Problem-solution

Overview

The topic for discussion as raised in the introductory paragraph can be developed in several patterns, such as problem-solution, classification, cause-effect, comparison and contrast. This unit explains how to organize a text in the pattern of problem-solution.

If a problem is raised in the introductory paragraph, then the body of the essay is usually focused on how to solve it, which is the problem-solution pattern of organization. Accordingly, this pattern of organization roughly falls into two types in terms of the nature of the issue, i.e., how to solve a problem or improve a situation and how to develop some skill / ability.

Type 1: How to analyze a problem for solutions

In this type, you will be required to describe a newsworthy current problem or bad situation facing the whole globe, a country or a locality, and propose one or more viable solutions. Examples include global issues like the shortage of energy, environmental pollution and global warming, national problems like unbalanced development in rural and urban areas or different parts of the country, and local problems like city congestion, too expensive housing and so on. Sometimes, however, an analysis of factors that cause the problem may serve as the hint for solutions.

Read the following passage to study how the writer analyzes the problem and what suggestion is given to address it.

For more than 600 years people have complained that youngsters cannot write proper English anymore. Even ancient Sumerian schoolmasters worried about the state of the "scribal art" in the world's first written language. Two universal truths emerge: Languages are always changing, and older people always worry that the young are not taking proper care of the language.

But what if the elders have a point? Of course language always changes, but could technology (or a simple increase in youthful lack of respect for tradition) mean that in some ages it changes faster than in others? Is change accelerating? In this case, a real problem could arise. Even if language change is not harmful, the faster language changes, the less new generations will be able to understand what their forebears wrote.

Maybe a greater conservatism would let modern readers peer further back in their own literary history. The problem is that conservatism works differently on writing than it does on speech. Writing is more permanent, so people choose their words more carefully and conservatively. It is slow and considered, so people can avoid new usages widely seen as mistakes. It is taught carefully by adults to children, which naturally exerts some conservative drag on the written language. And it is often edited, so the report of a young journalist with a casual contemporary style may well be edited to a more traditional one by an older editor.

Speech is different: Instead of permanent, slow, considered and taught, it is impermanent, fast, spontaneous and learned naturally by children from their surroundings. Speech will – at almost any level of linguistic conservatism – change faster than written language.

Written language only partly reflects speech. Younger writers introduce spoken or new words or usages into their writing, annoying their elders as they do. But no one dares write anything serious in phonetic spelling; English speakers are stuck with an old-fashioned and disordered system. Liberties with grammar – making the written language look like the spoken one – should be few and cautious. Giving the written language a little room to change, but not too much, is the only way to enjoy the best of both stability and vitality. The alternative – perfectly conservative writing – is a recipe for making writing less and less like the language future generations will speak, and thus less relevant to writing about the world they live in.

The problem	Fast changes in language make it difficult for youngsters to understand what their forebears wrote.
Analysis of the problem	Language in writing changes more slowly and remains more conservative than that used in speech.
Suggestion for solving the problem	Giving the written language a little room to change, but not too much, is the only way to enjoy the best of both stability and vitality.

1 Read the following passage and complete the table.

In the corporate world, digital defences are being defeated alarmingly often. As technology rapidly advances, many more things, from pacemakers to cars and power stations, are being connected to the Internet and governed by software that is vulnerable to criminals and terrorists.

No sooner is security strengthened than someone attempts to get around it. For instance, firms are experimenting with biometrics, or replacing passwords with things such as fingerprint or facial scans. But hackers are already looking for ways to steal copies of fingerprints – which, unlike passwords, cannot be changed easily – and to fool facial-recognition systems.

Innovations that generate increasing benefits are also making society more vulnerable. One example is the spread of sophisticated software, which has enabled companies to automate everything from stock-trading to credit-checking, dramatically lowering costs and prices. But digital Al Capones may target large numbers of people with automated "ransomware" attacks that lock them out of their computers and force them to pay a fee to get hold of their data again.

So how can companies better protect themselves in future? It is pointed out that much software code is being shipped with flaws or "bugs" in it that are "patched" over time. But hackers can exploit these bugs before they are remedied. One proposal is to strengthen liability laws so software companies that ship bug-ridden code can be sued more easily. This is worth debating, but a regime that was too harsh would force firms out of business and suppress innovation.

The problem	1) _____.
Analysis of the problem	• The software used 2) _____. • Hackers are active in developing or using 3) _____. Examples: 4) _____. 5) _____.
Suggestion for solving the problem	6) _____.
Comment on the suggestion	7) _____.

2 Read the following passage and complete the table.

Americans are used to thinking that law-and-order is threatened mainly by stereotypical violent crime. When the foundations of US law have actually been shaken, however, it has always been because ordinary law-abiding citizens took liberties with the law. Major instance: the *National Prohibition Act*. Recalls Donald Barr Chidsey in *On and Off the Wagon*: "Law breaking proved to be not painful, not even uncomfortable, but, in a mild and perfectly safe way, exciting." People wiped out the *Prohibition* at last not only because of the alcohol issue but because law breaking was seriously undermining the authority and legitimacy of government. Ironically, today's law breaking spirit, whatever its origins, is being encouraged unwillingly by government at many levels. The failure of police to enforce certain laws is only the surface of the problem: They take their order from the officials and constituents they serve. Worse, most state legislatures have helped destroy popular compliance with the federal 55 mph law, some of them by enacting puny fines that trivialize law-breaking activities. On a higher level, the

Administration in Washington has dramatized its wish to nullify civil rights laws simply by opposing instead of supporting certain court ordered desegregation rulings. With considerable justification, environmental groups, in the words of *Wilderness* magazine, accuse the Administration of "destroying environmental laws by failing to enforce them, or by enforcing them in ways that deliberately encourage noncompliance". Translation: law breaking at the top.

The most disquieting thing about the law-breaking spirit is its extreme infectiousness. Only a terminally foolish society would sit still and allow it to spread indefinitely.

The problem	1) _____.
Analysis of the problem	• Historically, ordinary citizens 2) _____. For example, the *National Prohibition Act* was wiped out. Today, the law-breaking spirit is 3) _____. • 4) _____. • 5) _____. • 6) _____.
Suggestion for solving the problem	7) _____.

Type 2: How to develop skills

As for this type, the topic may include how to develop social skills, how to get prepared for a job interview, how to promote a product, how to accommodate to a new place, etc.

Read the following passage, paying particular attention to the problem-solution pattern of its organization.

A major problem has been criticized about students' difficulty in using English despite years of study. About half of college students feel that they have made little or even no progress in English when they finish the college English courses. Teachers have remained the main targets for such criticism and complaints, blamed for their "low efficiency" and "boring method" of teaching and their inability of inspiring students.

But to me, things are not that simple. Investigations show that about 60% of students admit that they spend less than two hours of their spare time on English in a week. Many of such students explain that they simply cannot afford more time to study English because they have to do assignments in other subjects, which are much heavier than those in English.

Accordingly, more practice is needed if students really want to improve their English during their college life, which is a truth that can be denied by nobody, as a large number of top students have experienced. The real problem is, therefore, how to keep practicing English. On the part of teachers, more challenging but meaningful work or projects of English practice should be designed and assigned to students so that they can be kept busy with English. On the part of students, the most important point they need to keep in mind, or the most effective idea they need to practice, is that practice is the only (also effective, of course) key to better and proficient English. It is easy to imagine that one can do nothing well with English if he or she has a vocabulary of less than 5,000 words. Likewise, it is easy to say that only by keeping listening to English programs for at least 30 minutes a day can a learner improve his or her listening comprehension ability step by step. As to the speaking proficiency, no one can speak good English without reciting articles amounting to 10,000 words.

It should be noted that some high-sounding rhetoric has been harmful and misleading. One is the so-called "pleasure learning" or "learning through entertainment". While effective to children learners, this concept does not work so well with college students as adult learners of English. The other is the commercially preached "methods of learning" or even "secrets of learning", which induce students to be searching for such methods or secrets all day long but not learning. As adults, college students should be clear that if there is really such a magic way of learning under the sun at all, it is practice, nothing else.

The problem	**Phenomena** • College students have difficulty in using English despite years of study. • College students spend little time on English. **The problem (skill to be developed)** How to keep practicing English?
Solutions	**On the part of teachers** More challenging but meaningful work or projects of English practice should be designed and assigned to students so that they can be kept busy with English. **On the part of students** Practice is the key to better and proficient English. • To build a vocabulary of over 5,000 words • To listen to English programs for at least 30 minutes a day • To recite articles amounting to 10,000 words
Conclusion	Practice is the only magic method of learning English well.

3 Read the following passage and complete the table.

In a successful job interview, both the interviewer and the interviewee are trying to sell something. While the latter, i.e., the prospective employee, is trying to prove that he or she is the best candidate for the position advertised, the former, i.e., the company's representative, is not only seeing if the applicant will fit in, but also hoping to present a terrific workplace. Therefore, proper etiquette for both parties will help them achieve their goals respectively. Here are a few guidelines when it comes to job interview manners in the process of the interview.

Before the applicant arrives, the interviewer should take a few moments to get familiar with the applicant's résumé. Do not ask illegal questions or try to embarrass the candidate since this makes both you and your organization look unprofessional.

Applicants should always prepare in advance. Being well-groomed and choosing clean, professional clothing send the message that you have respect for the company, as well as for yourself. In addition, bring a few extra copies of your résumé and references with you in case that your résumé gets lost in the pile. By anticipating this possibility, you are showing concern for your potential employer.

The interviewer's time is valuable. Show that you recognize this by arriving early enough to be settled and prepared a few minutes before your appointment time. Be sure to extend your hand and introduce yourself when you meet. Don't forget to smile! Be friendly and open with the interviewer. Don't make it difficult for him or her by offering very little to the conversation. Ask relevant questions that show a genuine interest in the position and offer honest, complete responses to his or her questions. When the interview is complete, be sure to thank the interviewer for his or her time. Mention that you enjoyed meeting him or her and hope to be working together soon.

The problem (skill to be developed)	1) _____ is important to successful job interviews.
Skills for the interviewer	2) _____. 3) _____.
Skills for the interviewee	Getting prepared in advance: 4) _____. During the interview: 5) _____. At the end of the interview: 6) _____.

Conclusion

Both types of issues require the writer to propose feasible solutions, suggestions or proposals. What should be kept in mind is that the solutions offered should be specific and operational. In other words, they should not be too empty or general.

Writing assignment

Choose one of the topics below and compose a short essay of about 120 words. Use the pattern of problem-solution.

1. How to develop social skills?
2. How to reduce fossil fuel emissions?

Sharing

You live in an underdeveloped town, and the local government is holding a citizen forum on ideas to develop your town. Work in groups to make a proposal for the forum, focusing on how to improve gender equality and human health by upgrading energy access.

1 Work with your group members and discuss the following questions based on your prior knowledge and what you have learned in this unit.

1) What is the relationship between energy access and public health?
2) What roles does energy play in gender equality?
3) What can be done to improve the two issues by upgrading energy access?

2 Prepare a presentation with PPT including the three aspects discussed above.

3 Give your presentation in class.

Unit 3 Energy and Environment

▶ Lead-in

Fossil fuels are relatively cheap to use. However, it has been well recognized that they bring about serious consequences to the environment. The major environmental problems resulting from burning fossil fuels include global warming, air pollution, thermal pollution and maybe others. Can you explain how emissions bring about these impacts?

If nothing is done to the current situation of energy consumption, specifically, to solve fossil fuel dependency, the climate change will be catastrophic. The key solution lies in the fundamental transformation in our energy consumption behavior, to substantially reduce the consumption of fossil fuels, and to capture and bury the carbon emissions. What strategies do we have at hand?

▶ Learning objectives

Upon completion of this unit, you will be able to:
- talk about the harm done to health and environment by fossil fuel consumption;
- demonstrate the problem of global warming caused by carbon emissions and the solutions;
- use numbers properly and effectively in academic writing;
- recognize and use the pattern of comparison and contrast in writing;
- hold a panel discussion on the relation between fossil fuel consumption and environmental protection.

Listening and speaking

Listening

New words

prerequisite /priːˈrekwəzɪt/ n. 前提；必要条件
conservation /ˌkɒnsəˈveɪʃn/ n. 节约；保护
stringent /ˈstrɪndʒ(ə)nt/ a. 严格的
haze /heɪz/ n. 烟雾；霾

1 Listen to the recording and rank the concepts or policies in time order.

☐ The concept of green development was first discussed.
☐ Going green was recognized as the prerequisite of sustainable development.
☐ The strategic plan of energy conservation and emissions reduction was first proposed.

2 Listen to the recording again and fill in the blanks with what you hear.

1 Eco-provinces
The State Environmental Protection Administration, now the Ministry of Ecology and Environment approved a 1) _____ program in 2013 that introduced seven ecological provinces nationwide. At present, 2) _____ provinces, cities and villages cover almost half of the country.

2 Energy conservation and emissions reduction
The strategic plan of energy conservation and emissions reduction has led to remarkable achievements in 3) _____ energy consumption and greenhouse gas emissions.

3 Clean energy
Because of its huge population and the stage it is at in its development, China has 4) _____ much more to the development of clean energy. China has always strived to 5) _____ the costs of pollution.

4 The most stringent environmental protection law in China's history
Introduced in 2015 and seen as the 6) _____ law of its kind to date, China's revised Environmental Protection Law introduced 7) _____ punishments, promising to name and shame enterprises which break the law.

Energy and Environment | 57

5 Air pollution prevention

In northern China, home to Beijing, Tianjin and Hebei Province, a 8) _____ haze at times plagues the regions. Under the new Atmospheric Pollution Prevention and Control Law, which came into 9) _____ on January 1, 2016, ventilation corridors will be built in Beijing and outdated production 10) _____ will be closed down gradually.

Speaking

1 Read the short passage aloud and pay attention to the idea about car emissions and global warming.

In total, the US transportation sector – which includes cars, trucks, planes, trains, ships and freight – produces nearly 30% of all US global warming emissions, more than almost any other sector. The solutions are cleaner vehicles and cleaner fuels. Fuel efficient vehicles use less gas to travel the same distance as their less efficient counterparts. When we burn less fuel, we generate fewer emissions. When emissions go down, the pace of global warming slows.

Electric cars and trucks use electricity as fuel, producing fewer emissions than their conventional counterparts. When the electricity comes from renewable sources, all-electric vehicles produce zero emissions to drive.

Cleaner fuels produce fewer emissions when they're burned. Some fuels – such as those made from biofuels – can reduce emissions by 80% compared to gasoline.

2 Work in pairs and answer the two questions.

1 What are the solutions for global warming emissions in the US transportation sector?

_____

_____

2 What are the pros and cons of electric cars?

_____

_____

3 Do some research on global warming, a term you will find in the texts in this unit, and then prepare a one-minute oral presentation. Make sure the messages are properly delivered to the audience.

Extensive reading

Reading text

The Hidden Cost of Fossil Fuels

1. We've all paid a utility bill or purchased gasoline. Those represent the direct costs of fossil fuels: money paid out of pocket for energy from coal, natural gas and oil. But those expenses don't reflect the total cost of fossil fuels to each of us individually or to society as a whole. Known as externalities, the hidden costs of fossil fuels aren't represented in their market price, despite serious impacts on our health and environment.

2. Some of the most significant hidden costs of fossil fuels are from the air emissions that occur when they are burned. Unlike the extraction and transport stages, in which coal, oil and natural gas can have very different types of impacts, all fossil fuels emit carbon dioxide and other harmful air pollutants when burned. These emissions lead to a wide variety of public health and environmental costs that are borne at the local, regional, national and global levels.

Global warming emissions

3. Of the many environmental and public health risks associated with burning fossil fuels, the most serious in terms of its universal and potentially irreversible consequences is global warming. In 2014, approximately 78% of US global warming emissions were energy-related emissions of carbon dioxide. Of this, approximately 42% was from oil and other liquids, 32% from coal and 27% from natural gas.

4. Non-fossil fuel energy generation technologies, like wind, solar and geothermal, contributed less than 1% of the total energy-related global warming emissions. Even when considering the full life cycle carbon emissions of all energy sources, coal, oil and natural gas clearly stand out with significantly higher greenhouse gas emissions.

Air pollution

5. Burning fossil fuels emits a number of air pollutants that are harmful to both the environment and public health.

6 Sulfur dioxide (SO_2) emissions, primarily the result of burning coal, contribute to acid rain and the formation of harmful particulate matter. In addition, SO_2 emissions can exacerbate respiratory ailments, including asthma, nasal congestion and pulmonary inflammation. In 2014, fossil fuel combustion at power plants accounted for 64% of US SO_2 emissions.

7 Nitrogen oxides (NO_x) emissions, a byproduct of all fossil fuel combustion, contribute to acid rain and ground-level ozone (smog), which can burn lung tissue and can make people more susceptible to asthma, bronchitis and other chronic respiratory diseases. Fossil fuel-powered transportation is the primary contributor to US NO_x emissions.

8 Acid rain is formed when sulfur dioxide and nitrogen oxides mix with water, oxygen, and other chemicals in the atmosphere, leading to rain and other precipitation that is mildly acidic. Acidic precipitation increases the acidity of lakes and streams, which can be harmful to fish and other aquatic organisms. It can also damage trees and weaken forest ecosystems.

9 Particulate matter emissions produce haze and can cause chronic bronchitis, aggravated asthma and elevated occurrence of premature death. In 2010, it is estimated that fine particle pollution from US coal plants resulted in 13,200 deaths, 9,700 hospitalizations and 20,000 heart attacks. The impacts are particularly severe among the young, the elderly and those who suffer from respiratory disease. The total health cost was estimated to be more than $100 billion per year.

10 Mercury emissions to the air originate primarily from coal-fired power plants. As airborne mercury settles onto the ground, it washes into bodies of water where it accumulates in fish, and subsequently passes through the food chain to birds and other animals. The consumption of mercury-laden fish by pregnant women has been associated with neurological and neurobehavioral effects in infants. Young children are also at risk.

11 A number of studies have sought to quantify the health costs associated with fossil fuel-related air pollution. The National Academy of Sciences assessed the costs of SO_2, NO_x and particulate matter air pollution from coal and reported an annual cost of $62 billion for 2005 – approximately 3.2 cents per kilowatt hour (kWh). A separate study estimated that the pollution costs from coal combustion, including the effects of volatile organic compounds (VOCs) and ozone, were approximately $187 billion annually, or 9.3 cents per kWh.

12 A 2013 study also assessed the economic impacts of fossil fuel use, including illnesses, premature mortality, workdays lost, and direct costs to the healthcare system associated with emissions of particulates, NO_x, and SO_2. This study found

an average economic cost (or "public health added cost") of 32 cents per kWh for coal, 13 cents per kWh for oil and 2 cents per kWh for natural gas. While cost estimates vary depending on each study's scope and assumptions, together they demonstrate the significant and real economic costs that unpriced air emissions impose on society.

Thermal pollution

13 Power plants that return water to nearby rivers, lakes or the ocean can harm wildlife through what is known as "thermal pollution". Thermal pollution occurs due to the degradation of water quality resulting from changes in water temperature. Some power plants have large impacts on the temperature of nearby water sources, particularly coal plants with once-through cooling systems. For a typical 600-megawatt once-through system, 70 billion to 180 billion gallons of water cycle through the power plant before being released back into a nearby source. This water is much hotter (by up to 25°F) than when the water was initially withdrawn. Because this heated water contains lower levels of dissolved oxygen, its reintroduction to aquatic ecosystems can stress native wildlife, increasing heart rates in fish and decreasing fish fertility.

The future of energy

14 Burning coal, oil and natural gas has serious and long-standing negative impacts on public health, local communities and ecosystems, and the global climate. Yet the majority of fossil fuel impacts are far removed from the fuels and electricity we purchase, hidden within public and private health expenditures, military budgets, emergency relief funds, and the degradation of sensitive ecosystems. We don't pay for the cost of cancer, or the loss of fragile wetlands, when we pay our electricity bill – but the costs are real. Understanding these impacts is critical for evaluating the true cost of fossil fuels – and for informing our choices around the future of energy production.

Culture notes

nitrogen oxides (NO_x): It's most commonly used to refer to two species of oxides of nitrogen – nitric oxide (NO) and nitrogen dioxide (NO_2). In the atmosphere, nitrogen oxides mix with water vapor producing nitric acid. This acidic pollution can be transported by wind over many hundreds of miles, and deposited as acid rain.

ground-level ozone: Ozone occurs both in the earth's upper atmosphere and at ground level. It can be good or bad, depending on where it is found. Ozone at ground level is a harmful air pollutant, because of its effects on people and the environment, and it is the main ingredient in smog. It can trigger a variety of health problems, particularly for children, the elderly, and people of all ages who have lung diseases such as asthma.

volatile organic compounds (VOCs): They are organic chemicals that have a high vapor pressure at ordinary room temperature. Their high vapor pressure results from a low boiling point, which causes large numbers of molecules to evaporate or sublimate from the liquid or solid form of the compound and enter the surrounding air, a trait known as volatility.

Vocabulary

externality /ˌekstɜːˈnæləti/ n. 外部事物；外部效应
emit /ɪˈmɪt/ v. 排放；散发
pollutant /pəˈluːtnt/ n. 污染物
irreversible /ˌɪrɪˈvɜːsəbl/ a. 不可逆转的
life cycle n. 生命周期
sulfur /ˈsʌlfə/ n. 硫
exacerbate /ɪɡˈzæsəˌbeɪt/ v. 使加剧；使恶化
respiratory /rɪˈspɪrət(ə)ri; ˈresp(ə)rət(ə)ri/ a. 呼吸的
ailment /ˈeɪlmənt/ n. 小病；不适
asthma /ˈæsmə/ n. 哮喘病；气喘
nasal /ˈneɪzl/ a. 鼻的
congestion /kənˈdʒestʃ(ə)n/ n. 堵塞；拥塞
pulmonary /ˈpʌlmən(ə)ri/ a. 肺的
inflammation /ˌɪnfləˈmeɪʃn/ n. 发炎；炎症
nitrogen /ˈnaɪtrədʒ(ə)n/ n. 氮
oxide /ˈɒksaɪd/ n. 氧化物
byproduct /ˈbaɪˌprɒdʌkt/ n. 副产品；附带的结果
ozone /ˈəʊˌzəʊn/ n. 臭氧
susceptible /səˈseptəbl/ a. 易受影响的；易得病的
bronchitis /brɒŋˈkaɪtɪs/ n. 支气管炎

chronic /ˈkrɒnɪk/ a. 慢性的；长期的
precipitation /prɪˌsɪpɪˈteɪʃn/ n. 降水
acidic /əˈsɪdɪk/ a. 酸的；酸性的
aquatic /əˈkwætɪk/ a. 水生的；水栖的
ecosystem /ˈiːkəʊˌsɪstəm/ n. 生态系统
aggravated /ˈæɡrəˌveɪtɪd/ a. 加重的；严重的
elevated /ˈeləˌveɪtɪd/ n. 偏高的
occurrence /əˈkʌrəns/ n. 发生；出现
particle /ˈpɑːtɪkl/ n. 微粒
mercury /ˈmɜːkjʊri/ n. 汞
originate /əˈrɪdʒəˌneɪt/ v. 起源
airborne /ˈeəˌbɔːn/ a. 空气传播的；空气中的
neurological /ˌnjʊərəˈlɒdʒɪkl/ a. 神经病学的；神经系统的
neurobehavioral /ˌnjʊərəʊbɪˈheɪvjərəl/ a. 神经行为的
infant /ˈɪnfənt/ n. 婴儿；幼儿
quantify /ˈkwɒntɪˌfaɪ/ v. 量化；定量
kilowatt hour n. 千瓦时
volatile /ˈvɒləˌtaɪl/ a. 挥发性的
organic /ɔːˈɡænɪk/ a. 有机的
compound /ˈkɒmpaʊnd/ n. 化合物
mortality /mɔːˈtæləti/ n. 死亡数；死亡率
unpriced /ˌʌnˈpraɪst/ a. 未定价的；未标价的

impose /ɪmˈpəʊz/ v. 把…强加给
degradation /ˌdeɡrəˈdeɪʃn/ n. 退化；恶化
initially /ɪˈnɪʃli/ ad. 起初
dissolve /dɪˈzɒlv/ v. 溶解
fertility /fɜːˈtɪləti/ n. 繁殖力；生育能力

long-standing /ˌlɒŋˈstændɪŋ/ a. 长期存在的；长久持续的
expenditure /ɪkˈspendɪtʃə/ n. 支出；花费
wetlands /ˈwetlændz/ n. [pl.] 湿地；沼泽地

▶ Integrated exercises

1 Match each of the following statements with the paragraph from which the information is derived. You may choose a paragraph more than once.

_____ 1 Wind, solar and geothermal generation technologies contributed less than 1% of the total energy-related global warming emissions.

_____ 2 The impacts of particulate matter emissions are particularly severe among the young, the elderly and those who suffer from respiratory disease.

_____ 3 The heated water released from power plants contains lower levels of dissolved oxygen, so when returned to aquatic ecosystems it can stress native wildlife.

_____ 4 SO_2 emissions may aggravate respiratory ailments, including asthma, nasal congestion and pulmonary inflammation.

_____ 5 Pregnant women and young children are at risk from the consumption of mercury-laden fish.

_____ 6 It is very important for us to understand the hidden costs of fossil fuels in that it helps us make informed choices on energy.

_____ 7 Fossil fuel-powered transportation is the major contributor to US NO_x emissions.

_____ 8 Burning fossil fuels can generate air emissions with significant hidden costs.

_____ 9 Acidic precipitation increases the acidity of lakes and streams, which can do harm to fish and other aquatic organisms.

_____ 10 According to the findings of a study there is an average economic cost of 32 cents per kWh for coal.

2 Complete the following table to check your understanding of the major points and structure of the text.

Introduction (Paras. 1-2)	Burning fossil fuels involves 1) _____ not represented in our utility bills despite serious impacts on our health and environment.
Body (Paras. 3-13)	The hidden costs in burning fossil fuels are of three types. • **Global warming emissions** Compared with non-fossil fuel energy generation technologies, fossil fuels stand out with significantly higher 2) _____, leading to the universal and potentially irreversible consequence – 3) _____. • **Air pollution** Burning fossil fuels emits a number of 4) _____, which pose threat to both 5) _____ and 6) _____, and according to a number of studies, impose significant and real 7) _____ on society. • **Thermal pollution** Water used in the power plants, when released back to 8) _____, is much hotter and contains lower levels of 9) _____. Its reintroduction to aquatic ecosystems can 10) _____.
Conclusion (Para. 14)	Burning fossil fuels has serious and long-standing negative impacts on public health, local communities and 11) _____, and the global climate. Understanding these impacts is critical for evaluating 12) _____ – and for informing our choices around the future of energy production.

3 Complete the following table with the major types of air pollutants and their impacts on the environment and public health.

Air pollutants	Environmental costs	Public health costs
Sulfur dioxide	• 1) _____ • 2) _____	respiratory ailments, including: • 3) _____ • 4) _____ • 5) _____
Nitrogen oxides	• 6) _____ • 7) _____	burning lung tissue and making people more susceptible to: • 8) _____ • 9) _____ • other chronic respiratory diseases
Particulate matter	• 10) _____	• 11) _____ • 12) _____ • 13) _____
Mercury	• soil pollution • 14) _____ • disturbing ecosystems	threats to: • 15) _____ • young children

4 Work in pairs and take turns giving examples to demonstrate the harmful impacts of fossil fuel consumption on the environment. Take notes of the examples while you do this task.

_____
_____
_____

5 Translate the phrases into Chinese.

1 utility bill _____
2 direct cost _____
3 hidden cost _____
4 market price _____
5 harmful air pollutant _____
6 a wide variety of _____
7 public health _____
8 irreversible consequence _____
9 stand out _____
10 greenhouse gas emissions _____
11 military budget _____
12 acid rain _____
13 respiratory ailment _____
14 nasal congestion _____
15 pulmonary inflammation _____
16 aquatic organism _____
17 forest ecosystem _____
18 chronic bronchitis _____
19 premature mortality _____
20 coal-fired power plant _____

Energy and Environment | 65

21 food chain _____		23 healthcare system _____
22 at risk _____		24 thermal pollution _____

6 Complete the sentences by translating the Chinese given in brackets into English. Refer to the phrases listed above when necessary.

1 Dime stores specialize in _____ (种类繁多的便宜货).
2 Counterfeit medical products have long been _____ (对公共健康的威胁).
3 About half the electricity consumed in this country _____ (来自燃煤电厂).
4 As with all diseases, certain people _____ (比另一些人更容易受到威胁).
5 The common cold generally begins with _____ (流鼻涕、鼻塞、打喷嚏等症状).

Intensive reading

Warming up

1 We talk a lot about the greenhouse effect. Then what is a greenhouse and how does the greenhouse effect work? What is the relation between the greenhouse effect, global warming and climate change? Completing the following passage may help clarify these questions.

A greenhouse is a place to 1)_____, even in winter. Usually a greenhouse looks like a 2)_____ building. The glass lets in the radiant energy from 3)_____, but as the light passes through the glass, its wavelengths get longer and cannot pass easily back out of the glass. The radiant energy is trapped as 4)_____.

Some 5)_____ in the atmosphere trap heat in much the same way as a layer of glass. They are called greenhouse gases. Sunlight enters the earth's atmosphere. Some of the energy reflects back, but some is trapped in the 6)_____ by greenhouse gases, causing temperatures to 7)_____. That is how 8)_____ works.

9)_____ is a planet-wide rise in temperature. Rising temperatures may cause changes in rainfall and the strength of storms, 10)_____ polar ice and 11)_____ sea levels. This is called 12)_____.

2 Work in groups and discuss the greenhouse gases you know and where they come from.

Reading text

Global Warming: It's About Energy

1   Finally, after years of effort by dedicated scientists and activists like Al Gore, the issue of global warming has begun to receive the international attention it desperately needs. The publication on February 2 of the most recent report

by the Intergovernmental Panel on Climate Change (IPCC), providing the most persuasive evidence to date of human responsibility for rising world temperatures, generated banner headlines around the world. But while there is a growing consensus on humanity's responsibility for global warming, policymakers have yet to come to terms with its principal cause: our unrelenting consumption of fossil fuels (primarily coal, fuel oil and natural gas).

2 When talk of global warming is introduced into the public discourse, as in Gore's *An Inconvenient Truth*, it is generally characterized as an environmental problem, akin to water pollution, air pollution, pesticide abuse, and so on. This implies that it can be addressed – like those other problems – through a concerted effort to "clean up" our resource-utilization behavior, by substituting "green" products for ordinary ones, by restricting the release of toxic substances, and so on.

3 But global warming is not an "environmental" problem in the same sense as these others – it is an energy problem, first and foremost. Almost 90% of the world's energy is supplied through the combustion of fossil fuels, and every time we burn these fuels to make energy we release carbon dioxide into the atmosphere; carbon dioxide, in turn, is the principal component of the "greenhouse gases" that are responsible for warming the planet. Energy use and climate change are two sides of the same coin.

Fossil fuel dependency

4 Consider the situation in the United States. According to the Department of Energy (DoE), carbon dioxide emissions constitute 84% of this nation's greenhouse gas emissions. Of all US carbon dioxide emissions, most – 98% – are emitted as a result of the combustion of fossil fuels, which currently provide approximately 86% of America's total energy supply. This means that energy use and carbon dioxide emissions are highly correlated: The more energy we consume, the more CO_2 we release into the atmosphere, and the more we contribute to the buildup of greenhouse gases.

5 Because Americans show no inclination to reduce their consumption of fossil fuels – but rather are using more and more of them all the time – one can foresee no future reduction in US emissions of greenhouse gases. According to the DoE, the United States is projected to consume 35% more oil, coal and gas combined in 2030 than in 2004; not surprisingly, the nation's emissions of carbon dioxide are expected to rise by approximately the same percentage over this period. If these projections prove accurate, total US carbon dioxide emissions in 2030 will reach a staggering 8.1 billion metric tons, of which 42% will be generated through the consumption of oil (most of it in automobiles, vans, trucks and buses), 40% by the burning of coal (principally to produce electricity), and the

remainder by the combustion of natural gas (mainly for home heating and electricity generation). No other activity in the United States will come even close in terms of generating greenhouse gas emissions.

6 What is true of the United States is also true of other industrialized and industrializing nations, including China and India. Although a few may rely on nuclear power or renewables to a greater extent than the United States, all continue to consume fossil fuels and to emit large quantities of carbon dioxide, and so all are contributing to the acceleration of global climate change. According to the DoE, global emissions of carbon dioxide are projected to increase by a frightening 75% between 2003 and 2030, from 25.0 billion to 43.7 billion metric tons. People may talk about slowing the rate of climate change, but if these figures prove accurate, the climate will be much hotter in coming decades and this will produce the most damaging effects predicted by the IPCC.

7 What this tells us is that the global warming problem cannot be separated from the energy problem. If the human community continues to consume more fossil fuels to generate more energy, it inevitably will increase the emission of carbon dioxide and so hasten the buildup of greenhouse gases in the atmosphere, thus causing irreversible climate change. Whatever we do on the margins to ameliorate this process – such as planting trees to absorb some of the carbon emissions or slowing the rate of deforestation – will have only negligible effect so long as the central problem of fossil fuel consumption is left unchecked.

What to do

8 If, however, we seek to protect the climate while there is still time to do so, we must embrace a fundamental transformation in our energy behavior: Nothing else will make a significant difference. In practice, this devolves into two fundamental postulates. We must substantially reduce our consumption of fossil fuels, and we must find ways to capture and bury the carbon byproducts of the fossil fuels we do consume.

9 Various strategies have been proposed to achieve these objectives. Those that offer significant promise should be utilized to the maximum extent possible. This is not the place to evaluate these strategies in detail, except to make a few broad comments.

10 First, as noted, since 42% of American carbon dioxide emissions (the largest share) are produced through the combustion of petroleum, anything that reduces oil consumption – higher fuel efficiency standards for motor vehicles, bigger incentives for hybrids, greater use of ethanol, improved public transportation, carpooling, and so on – should be made a major priority.

11 Second, because the combustion of coal in electrical power plants is our next

biggest source of CO_2, improving the efficiency of these plants and filtering out the harmful emissions have to be another top priority.

12 Finally, we should accelerate research into promising new techniques for the capture and "sequestration" of carbon during the combustion of fossil fuels in electricity generation. Some of these plans call for burying excess carbon in hollowed-out coalmines and oil wells – a very practical use for these abandoned relics, but only if it can be demonstrated that none of the carbon will leak back into the atmosphere, adding to the buildup of greenhouse gases.

13 Global warming is an energy problem, and we cannot have both an increase in conventional fossil fuel use and a habitable planet. It's one or the other. We must devise a future energy path that will meet our basic (not profligate) energy needs and also rescue the climate while there's still time. The technology to do so is potentially available to us, but only if we make the decision to develop it swiftly and on a very large scale.

Culture notes

Al Gore: Albert Arnold Gore Jr. (born March 31, 1948) is an American politician and environmentalist who served as the 45th vice president of the United States from 1993 to 2001. After his term as vice president ended in 2001, Gore remained prominent as an author and environmental activist, whose work in climate change activism earned him (jointly with the IPCC) the Nobel Peace Prize in 2007.

the Intergovernmental Panel on Climate Change (IPCC): It is an intergovernmental body of the United Nations, dedicated to providing the policymakers with regular scientific assessments on climate change, its implications and potential future risks, and putting forward adaptation and mitigation options.

***An Inconvenient Truth*:** It is one of the best-selling books written by Al Gore. The book exposes the compelling case that global warming is real and man-made, and its effects will be cataclysmic if we don't act now. Its adaptation, the documentary film *An Inconvenient Truth*, has won an Academy Award for Best Documentary.

Vocabulary

activist /ˈæktɪvɪst/ n. 积极分子
desperately /ˈdesp(ə)rətli/ ad. 极度地；非常
publication /ˌpʌblɪˈkeɪʃn/ n. 出版
intergovernmental /ˌɪntəˌɡʌv(ə)nˈmentl/ a. 政府与政府间的
persuasive /pəˈsweɪsɪv/ a. 有说服力的
banner headline n. (报纸头版的) 通栏标题
policymaker /ˈpɒləsiˌmeɪkə/ n. 政策制定者
principal /ˈprɪnsəpl/ a. 主要的；最重要的
unrelenting /ˌʌnrɪˈlentɪŋ/ a. 持续的；不断的
discourse /ˈdɪskɔːs/ n. 话语；论述
pesticide /ˈpestɪˌsaɪd/ n. 杀虫剂
restrict /rɪˈstrɪkt/ v. 限制；控制
toxic /ˈtɒksɪk/ a. 有毒的
substance /ˈsʌbstəns/ n. 物质
first and foremost 首先；首要的事
dependency /dɪˈpendənsi/ n. 依赖
constitute /ˈkɒnstɪˌtjuːt/ v. 组成；构成
correlate /ˈkɒrəˌleɪt/ v. 使相互关联
buildup /ˈbɪldʌp/ n. 增长
inclination /ˌɪnklɪˈneɪʃn/ n. 倾向；意向
project /prəˈdʒekt/ v. 预计；推断
surprisingly /səˈpraɪzɪŋli/ ad. 惊人地；出人意料地
projection /prəˈdʒekʃn/ n. 预测；预计
staggering /ˈstæɡərɪŋ/ a. 大得惊人的；令人吃惊的
van /væn/ n. 厢式小型货车

remainder /rɪˈmeɪndə/ n. 剩余部分；剩余物
acceleration /əkˌseləˈreɪʃn/ n. 加快；增速
hasten /ˈheɪsn/ v. 加速
on the margins 处于边缘地位；处于非主体地位
ameliorate /əˈmiːliəˌreɪt/ v. 改善；改进
deforestation /diːˌfɒrɪˈsteɪʃn/ n. 砍伐树林
negligible /ˈneɡlɪdʒəbl/ a. 微不足道的
unchecked /ʌnˈtʃekt/ a. 未受抑制的；未受制止的
embrace /ɪmˈbreɪs/ v. 欣然接受；乐意采纳
devolve /dɪˈvɒlv/ v. 被转移；被移交
postulate /ˈpɒstjuˌleɪt/ n. 假设；假定
substantially /səbˈstænʃli/ ad. 大量地；可观地
ethanol /ˈeθəˌnɒl/ n. 乙醇
priority /praɪˈɒrəti/ n. 优先处理的事；当务之急
filter /ˈfɪltə/ v. 过滤；滤除
sequestration /ˌsiːkwəˈstreɪʃn/ n. 封存；隔离
excess /ˈekses/ a. 过多的；多余的
hollowed-out /ˈhɒləʊdˌaʊt/ a. 挖空的
coalmine /ˈkəʊlˌmaɪn/ n. 煤矿
relic /ˈrelɪk/ n. 残留物；遗迹
habitable /ˈhæbɪtəbl/ a. 适于居住的
devise /dɪˈvaɪz/ v. 设计；想出；发明
profligate /ˈprɒflɪɡət/ a. 挥霍的；浪费的

▶ Text understanding

1 What is the nature of global warming? Read Paras. 1-3 and complete the summary of these paragraphs.

Global warming is not 1) _____ in the same sense as others. Rather it is 2) _____ because of our excessive dependence on 3) _____, the burning of which releases 4) _____, the chief cause of global warming. Energy use and climate change are 5) _____.

2 What examples are cited to illustrate our heavy dependence on fossil fuels? Read Paras. 4-7 and complete the following table.

Examples	Facts & statistics
the United States	• Carbon dioxide emissions constitute 1) _____ of this nation's greenhouse gas emissions. 2) _____ of all carbon dioxide emissions come from the combustion of fossil fuels. • Fossil fuels provide approximately 3) _____ of America's total energy supply. • Carbon dioxide emissions in 2030 will reach a staggering 4) _____ metric tons, of which 42% will come from oil consumption, 40% by coal burning and the remainder by the combustion of natural gas.
Other nations, including China and India	• Despite an attempted gradual transition to 5) _____ in a few countries, all continue to consume fossil fuels and to emit large quantities of carbon dioxide. • Global emissions of carbon dioxide are projected to increase by a frightening 6) _____ between 2003 and 2030, from 25.0 billion to 43.7 billion metric tons.

3 What suggestions are given to cope with the issue of global warming? Read Paras. 8-13 and complete the following list.

1 _____

2 _____

3 _____

▶ Language building

1 Match each of the words in the left column with its corresponding meaning in the right column.

_____	1 discourse	a	remaining people, things or time; the rest
_____	2 consensus	b	a formal discussion of a topic in speech or writing
_____	3 akin	c	similar to sth.
_____	4 remainder	d	to have a mutual relationship or connection
_____	5 buildup	e	to be a distinctive trait or mark of sth.
_____	6 correlate	f	more than is necessary

_____ 7 staggering   g a gradual accumulation or increase of sth.
_____ 8 panel   h a generally accepted opinion
_____ 9 excess   i extremely surprising
_____ 10 characterize j a group of specialists giving advice or opinion on a particular subject

2 Among the three choices given, choose the one that is **NOT** close in meaning to the underlined word in each sentence.

1 If the human community continues to consume more fossil fuels to <u>generate</u> more energy, it inevitably will increase the emission of carbon dioxide.
   A) occur   B) create   C) produce

2 Policymakers have yet to come to terms with its principal cause: our <u>unrelenting</u> consumption of fossil fuels.
   A) ceaseless   B) conventional   C) constant

3 Carbon dioxide, in turn, is the <u>principal</u> component of the "greenhouse gases" that are responsible for warming the planet.
   A) chief   B) urgent   C) major

4 This implies that it can be addressed by restricting the release of <u>toxic</u> substances.
   A) harmless   B) lethal   C) poisonous

5 According to the Department of Energy (DoE), carbon dioxide emissions <u>constitute</u> 84% of this nation's greenhouse gas emissions.
   A) make up   B) comprise   C) compromise

6 Anything that reduces oil consumption – bigger <u>incentives</u> for hybrids, greater use of ethanol, and so on – should be made a major priority.
   A) acceleration   B) stimulus   C) encouragements

7 If, however, we seek to protect the climate while there is still time to do so, we must <u>embrace</u> a fundamental transformation in our energy behavior.
   A) support   B) include   C) welcome

8 Whatever we do on the margins to ameliorate this process will have only <u>negligible</u> effect so long as the central problem is left unchecked.
   A) insignificant   B) considerable   C) slight

9 It <u>inevitably</u> will increase the emission of carbon dioxide and so hasten the buildup of greenhouse gases in the atmosphere.
   A) unavoidably   B) certainly   C) accidentally

10 One can foresee no future <u>reduction</u> in US emissions of greenhouse gases.
   A) decline   B) cutback   C) substitution

3 Complete each of the following sentences with an appropriate word from the given word family. Change the form where necessary.

1 It took a lot of hard work and _____, but we managed to finish the project on time. (dedicate, dedication, dedicator, dedicatory)

Energy and Environment | 73

2 The park is open to the public without _____. (restrict, restrictive, restriction)

3 Students should aim to become more _____ of their teachers. (dependency, dependent, depend, independent)

4 His funeral will be private, in _____ with his wishes. (accord, according, accordance, accordingly)

5 She's _____ to gossip with complete strangers. (inclined, incline, inclination)

6 In general, I have been quite skeptical about the Internet's supposedly _____ effect on politics. (transform, transformation, transformative, transformable, transformer)

7 There is a certain _____ that e-book sales have now overtaken paperback sales on Amazon's US site. (inevitable, inevitably, inevitability)

8 This feeling of _____ and helplessness was common to most of the refugees. (desperation, desperate, desperately)

9 The unemployment rate has been _____ to fall. (project, projected, projection)

10 One percent of Brazil's total forest cover is being lost every year due to _____. (forest, deforest, reforest, deforestation)

4 Match each word in the box with the group of phrases where it is usually found.

| strategy | concerted | devise | priority | generate | consensus |

1 _____
a ~ effort
take ~ action
make a ~ attack
launch a ~ campaign

2 _____
a growing ~
a general ~
a lack of ~
reach a ~

3 _____
evaluate a ~
a long-term ~
a global marketing ~
develop a ~

4 _____
~ a path
~ a method
~ a new plan
~ a scheme

5 _____
a top ~
a first ~
take ~ over
give ~ to

6 _____
~ emissions
~ energy
~ electricity
~ power

5 Find the idiomatic expressions in the text matching the Chinese equivalents.

1 迄今为止 ＿＿＿＿＿
2 农药滥用 ＿＿＿＿＿
3 同心协力 ＿＿＿＿＿
4 清理整顿 ＿＿＿＿＿
5 有毒物质 ＿＿＿＿＿
6 首先 ＿＿＿＿＿
7 相应地 ＿＿＿＿＿
8 家庭取暖 ＿＿＿＿＿
9 从…方面说来 ＿＿＿＿＿
10 适用于 ＿＿＿＿＿
11 根本的转变 ＿＿＿＿＿
12 有重大影响 ＿＿＿＿＿
13 尽最大可能 ＿＿＿＿＿
14 概括性的评论 ＿＿＿＿＿
15 公共交通 ＿＿＿＿＿
16 滤除 ＿＿＿＿＿
17 头等大事 ＿＿＿＿＿
18 需要 ＿＿＿＿＿
19 挖空的煤矿 ＿＿＿＿＿
20 增加 ＿＿＿＿＿
21 大规模地 ＿＿＿＿＿

6 Combine the sentences given below. Then compare your sentences with the original ones in the text.

1 ＿＿＿＿＿＿＿＿＿＿＿＿＿＿＿＿＿＿＿＿＿＿＿＿＿＿＿＿＿＿＿＿＿＿＿

    a Scientists and activists like Al Gore have been dedicated to the issue of global warming and put years of effort into it.

    b Finally, the issue of global warming has begun to receive the international attention.

    c The issue of global warming needs the international attention desperately.

2 ＿＿＿＿＿＿＿＿＿＿＿＿＿＿＿＿＿＿＿＿＿＿＿＿＿＿＿＿＿＿＿＿＿＿＿

    a The IPCC published the most recent report on February 2.

    b The report provided the most persuasive evidence to date of human responsibility for rising world temperatures.

    c The publication generated banner headlines around the world.

3 ＿＿＿＿＿＿＿＿＿＿＿＿＿＿＿＿＿＿＿＿＿＿＿＿＿＿＿＿＿＿＿＿＿＿＿

    a Of all US carbon dioxide emissions, most – 98% – are emitted as a result of the combustion of fossil fuels.

    b Fossil fuels currently provide approximately 86% of America' total energy supply.

4 ＿＿＿＿＿＿＿＿＿＿＿＿＿＿＿＿＿＿＿＿＿＿＿＿＿＿＿＿＿＿＿＿＿＿＿

    a If these projections prove accurate, total US carbon dioxide emissions in 2030 will reach a staggering 8.1 billion metric tons.

    b Forty-two percent of the carbon dioxide emissions will be generated through the consumption of oil.

    c Forty percent of the carbon dioxide emissions will be generated by the burning of coal.

    d The remainder will be generated by the combustion of natural gas.

5 _____

   a We must devise a future energy path.
   b The path will meet our basic energy needs.
   c The path will also rescue the climate while there's still time.

7 Translate each of the Chinese sentences into English by using the underlined phrase or structure in the example.

1 Policymakers have yet to <u>come to terms with</u> its principal cause: our unrelenting consumption of fossil fuels.
她曾经给予我和家人莫大的帮助，她的死讯令人难以接受。
_____

2 It can be addressed by <u>substituting</u> "green" products <u>for</u> ordinary ones.
我们这把破损的椅子得换个新的。
_____

3 Americans <u>show</u> no <u>inclination to</u> reduce their consumption of fossil fuels.
他一开始就表现出我行我素的倾向。
_____

4 No other activity in the United States will <u>come</u> even <u>close in terms of</u> generating greenhouse gas emissions.
这些都是极好的歌曲，但是就销量而言没有一个能与《地球之歌》(Earth Song) 相比。
_____

5 What <u>is true of</u> the United States is also true of other industrialized and industrializing nations.
财富就像海水，越喝越渴；名声也是如此。
_____

6 We must embrace a fundamental transformation in our energy behavior: Nothing else will <u>make a</u> significant <u>difference</u>.
发挥重要作用的往往是细微之处。
_____

7 The technology to do so is potentially available to us, but <u>only if</u> we make the decision to develop it swiftly and on a very large scale.
我们全队只有人人各尽所能才能取得成功。
_____

Academic writing

▶ Micro-skill: Numbers

In academic writing, statistical data like numbers and percentages should be clearly and accurately stated.

1 The use of numbers

In introductions, an accurate account of a situation can be given with numbers. For example:

China has the world's largest number of terrestrial vertebrates, with a total of 1,800 species, which accounts for approximately one-tenth of the same species of vertebrates in the world.

The Wall Street Journal uses 220,000 metric tons of newsprint each year.

These vegetables are stored in a warehouse at temperatures below 20°C.

In whole numbers, there is no final "s" on hundred / thousand / million. For example:

Three thousand people study there.

Thousands of people were gathering in the square.

Note:
Normally, whole numbers should be written as words from one to ten and as digits above ten. For example:

At first, there were only two schools in the city, but now the number has risen to 14.

Other exact statistical measures are as follows.

Percentages	75% or 75 percent
Fractions	2/3 or two-thirds
Decimals	0.567
Currencies	$443 m or 443 million dollars

Exact statistical measures, such as percentages, decimals and mathematical operations, should always be written in numeral form. But use words for common fractions.

Instead of specific statistical data, the words *figures* and *numbers* can both be used in a general sense. For example:

The figures / numbers in the report need to be analyzed critically.

1 Complete the following sentences using the data in the table.

Year	2007	2008	2009	2010
Number of migrate workers	100	150	300	700

1 In the shoe factory, between 2007 and 2008, the number of migrate workers increased by _____ percent.
2 The number grew by _____ percent the following year.
3 Between 2007 and 2010 there was a _____ percent increase.

2 The simplification of numbers

Sometimes, instead of giving too many numbers, words such as *few*, *various*, *dozens* may be used if the actual number is not important. For example:

The economic crisis has forced 44 banks to close.
The economic crisis has forced many banks to close.

The following expressions can be used to simplify numbers.

few	almost none
a few	around five
several	three or four
various	more than four, less than seven
dozens of	more than 24, less than 60
scores of	more than 40, less than 100

2 Rewrite the following sentences using the expressions above to simplify the numbers.

1 Seventy-six students participated in the English contest.

2 Tom rewrote the letter three times.

3 This year 37 books will be published on new energy.

4 Only three people agreed to lend him some money.

5 She enrolled at Harvard University and stayed there for four years.

3 Numerical phrases

In academic writing, numerical phrases instead of numbers are often used to present statistics. For example, instead of referring to the rise from 34,000 to 70,000, the word *double* could be used.

The number of students going abroad to study roughly doubled in two years.

More numerical phrases include *one in / out of four, a tenfold increase, a quarter of, a small / large proportion, the majority / minority, the average*, etc. For example:

One in / out of four foreign students is from America.
There was a fivefold increase in the number of students.
A quarter of the employees are women.
The majority of the courses are taught in English.
A small proportion of the companies are making money.

In terms of percentage, the following numerical phrases can be used.
5%-20% = a tiny / small minority
21%-39% = a minority
40%-49% = a substantial / significant minority
51%-55% = a small majority
56%-79% = a majority
80% + = a large majority

3 Rewrite the following sentences in a simpler way, using suitable numerical expressions.

1 Researchers used to estimate that 2%-16% of the US population suffered compulsive buying disorder.

2 Out of 18 members in the research group 12 were Chinese.

3 The new subway reduced the commute from two hours to 55 minutes.

4 The number of students applying for the computer course has risen from 200 last year to 300 this year.

5 More than 80% of American students complete their PhD course; in Spain the figure is just 35%.

6   The total number of overweight and obese Americans increased. In 2001, nearly 400,000 Americans are either overweight or obese, in 2002, 820,000 and 1,710,000 in 2003.

7   An incandescent lamp costs $1.45 while a fluorescent lamp costs, on average, $2.87.

4 The following data are about 15 books published by a publishing house. Write five sentences about the books using the data.

Languages		Genres		Themes	
Arabic	2	Adventure	1	Love	2
Chinese	8	Classic	3	Travel	1
French	1	Memoir	2	Food	3
Japanese	1	Science fiction	6	Sports	3
Korean	2	Textbook	2	War	5
Spanish	1	Poetry	1	Business	1

1   A small majority of the books are published in Chinese.

2   _____

3   _____

4   _____

5   _____

6   _____

▶ Macro-skill: Comparison and contrast

Comparison and contrast are used when two or more things belonging to the same category are to be described or explained or when one certain thing that changes or develops over time needs to be examined. A comparison between objects obviously not in the same category, like an insect and a smartphone, may be of no significance or possibility. On the other hand, there is no need to compare two things that are identical. In other words, comparison and contrast are often applied to describe two similar things to reveal their significant differences, two seemingly unrelated things to discover their similarities so far unsuspected, or something that has become somewhat different over time. Common examples or topics for comparison and contrast can be "people's life: present and past" "table manners in China and in the West" "differences of life in the middle school and in the university" and so on.

Strictly speaking, comparison essays tend to focus on the similarities while contrast essays place emphasis on the differences between two or more items. In practice, however, comparison and contrast often go together because people tend to compare two things that are similar in certain aspects and different in others.

In various English tests, topics related to elaboration of advantages and disadvantages of something often call for the use of comparison and contrast, for example, shopping online and in physical stores, MOOCs and traditional classroom teaching, and a comparison between online (virtual) and face-to-face communication.

Sometimes a subject that is familiar to the reader may be used to explain another subject that is unfamiliar. If the two subjects are things belonging to different categories, then analogy, a special form of comparison, is used.

1 Read the following passage and complete the table.

Over the last 25 years, British society has changed a great deal – or at least many parts of it have. In some ways, however, very little has changed, particularly where attitudes are concerned. Ideas about social class – whether a person is "working class" or "middle class" – are one area in which changes have been extremely slow.

In the past, the working class tended to be paid less than middle-class people, such as teachers and doctors. As a result of this and also of the fact that workers' jobs were generally much less secure, distinct differences in lifestyles and attitudes came into existence. The typical working man would collect his wages on Friday evening and then, it was widely believed, having given his wife her "housekeeping", would go out and squander the rest on beer and betting.

The stereotype of what a middle-class man did with his money was perhaps nearer the truth. He was – and still is – inclined to take a longer-term view. Not only did he regard buying a house as a top priority, but he also considered the education of his children as extremely important. Both of these provided him and his family with security. Only in very few cases did workers have the opportunity (or the education and training) to make such long-term plans.

Nowadays, a great deal has changed. In a large number of cases factory workers earn as much as, if not more than, their middle-class supervisors. Social security and laws to improve job security, combined with a great rise in the standard of living since the mid-50s of the 20th century, have made it less necessary than before to worry about "tomorrow". Working-class people seem slowly to be losing the feeling of inferiority they had in the past. In fact there has been a growing tendency in the past few years for the middle classes to feel slightly ashamed of their position.

The subject to be compared		1) _____ in the past and at present
Comparison	In the past	Income: 2) _____.
		Job security: 3) _____.
		Lifestyle and attitude: The working class spent wages on 4) _____, while the middle class take 5) _____ and make 6) _____ for housing and children education.
	At present	Income: 7) _____.
		Job security and lifestyle: Thanks to a great rise in the standard of living and improvement of job security, workers 8) _____.
		Attitude: The working class no longer 9) _____, while the middle class tend to feel ashamed of their position.

2 Read the excerpt from *How to grow old* by Bertrand Russell and complete the table.

Some old people are oppressed by the fear of death. ... The best way to overcome it – so at least it seems to me – is to make your interests gradually wider and more impersonal, until bit by bit the walls of the ego recede, and your life becomes increasingly merged in the universal life. An individual human existence should be like a river – small at first, narrowly contained within its banks, and rushing passionately past boulders and over waterfalls. Gradually the river grows wider, the banks recede, the waters flow more quietly, and in the end, without any visible break, they become merged in the sea, and painlessly lose their individual being. The man who, in old age, can see his life in this way, will not suffer from the fear of death, since the things he cares for will continue.

Topic	1) _____.
Analogy	An individual human existence should be like 2) _____. • 3) _____. • 4) _____.
Conclusion	5) _____.

3 Read the following passage and complete the table.

The days in college and in high school are both important stages in my growth. Since I entered college a year ago, I have realized that there are some similarities as well as differences between college life and high school life.

The similarities exist in a number of ways. The major one is that learning or study is of paramount importance and occupies most of my time both in college and in high school. In other words, one has to work hard for good academic performance at both stages of life. Another similarity may be that I can make a lot of friends both in college and in high school.

However, differences may exceed similarities. To begin with, my spare time is spent in very different ways. In high school, I was obsessed with homework in various subjects, examinations and after-school training classes. But in college, I can be much more free in arranging and spending my spare time. Besides reading and writing assignments, I can go to various extracurricular activities, develop various skills, and do whatever I like (never do anything illegal of course). Another important difference is that college life sets higher demands for personal independence and self-management. University teachers are not so watchful as high school teachers in taking care of everything in our life. Parents stay back at home far away from us. All this requires us to manage our time, our money, and ultimately, our life, independently. For example, every semester I need to allocate the money to proper portions within budget on food, entertainment, groceries, clothing, sports, social activities and study.

In a word, in spite of the similarities between college life and high school life, there are differences in more aspects, such as spending one's spare time and developing the ability of self-management. While the similarities require us to never forget our original aspirations, the differences require us to make better use of our college time so that we can grow into better people in the future.

The subject to be compared		1) _____
Comparison	Similarities	2) _____.
		3) _____.
	Differences	4) _____.
		5) _____.
Conclusion		6) _____.

---- **Writing assignment** ----

Choose one of the topics below and compose a short essay of about 120 words. Use the pattern of comparison and contrast.

1. Illustrate the advantages and disadvantages of fossil fuels and renewable energy.

2. Analyze the influence of electric cars, both positive and negative, and then give your comment on it.

Sharing

> The International Environment Protection Conference is holding a panel discussion, in which experts in relevant fields discuss environmental issues. You are going to engage in the discussion and the topic is the relationship between fossil fuel consumption and environmental protection.

 Work in groups of five and choose your roles in the panel discussion.

Role	Responsibility
Host	Beginning and ending the discussion, introducing the speakers and involving the audience
Expert on energy	Introducing different forms of energy resources including fossil fuels and renewable energy resources, and their emissions
Expert on global warming	Explaining the impact of fossil fuel consumption on the earth's climate system, using examples for illustration
Expert on pollution	Explaining the impact of fossil fuel consumption on air and water, using examples for illustration
Secretary of IPCC	Proposing strategies for countries, businesses and individuals to reduce fossil fuel consumption to protect the earth

Prepare your individual speech with PPT according to your roles. Collect relevant information. You can refer to the texts in this unit or search on the Internet.

Hold the panel discussion using the following procedure.

1) The host introduces the panel and each participant;
2) The panelists deliver their presentation individually;
3) The host asks the panelists questions;
4) The host collects questions from the audience;
5) The host thanks everyone involved and makes a conclusion.

Unit 4 Energy and Future

▶ Lead-in

The energy industry is always looking for new ways to help fuel the future. Countries and areas are harnessing new technologies and techniques to meet their energy needs. It's hard to tell exactly what the future of energy is. Since abundant shale gas was discovered and extracted in the US, it is projected that the US will become energy independent in the near future, and the impact does not end within the country. What domestic and international influence will there be?

Another important change in the energy industry is the gradual popularization of renewables, among which the drop in the price of solar panels stimulates a surge in purchase and installation. Many advantages are evident in installing residential solar panels besides the free availability of sunlight and zero greenhouse emissions. Do you know what the other benefits are?

▶ Learning objectives

Upon completion of this unit, you will be able to:
- discuss the future of energy in the US and its impacts;
- explain the benefits of installing residential solar panels;
- use references and quotations correctly and properly in academic writing;
- recognize and use the pattern of cause-effect in writing;
- debate on the prediction of the US energy future.

Listening and speaking

Listening

New words

hydrogen /ˈhaɪdrədʒən/ n. 氢
squash /skwɒʃ/ v. 挤进；塞进
molecule /ˈmɒlɪˌkjuːl/ n. 分子
insulate /ˈɪnsjʊˌleɪt/ v. 使隔热

sweatbox /ˈswetbɒks/ n. 发汗装置
infrared /ˌɪnfrəˈred/ a. 红外线的
paddy field n. 稻田
methane /ˈmiːˌθeɪn/ n. 甲烷；沼气

1 Listen to the recording and decide whether the statements are true (T) or false (F).

☐ 1 The weight of exhaust gases from your car is the same as the weight of petrol burned.

☐ 2 The natural greenhouse effect is a good thing and keeps the earth warm enough for life to exist.

☐ 3 Burning wood has enhanced the greenhouse effect even though trees cut down are replanted.

☐ 4 Burning fossil fuels containing carbon that has remained underground adds greenhouse gases to the atmosphere.

2 Listen to the recording again and fill in the blanks with what you hear.

Before the world became industrialized by burning fossil fuels, carbon dioxide
1) _____ in the atmosphere was enough to keep us warm. Without this
natural 2) _____ of insulating gas, the earth will be too cold to support life.
But this status quo is starting to change. As 3) _____ adds carbon dioxide
into our atmosphere, it is as if we're putting a sweater around the planet. Some people
think that living things contribute to the 4) _____ greenhouse effect, because
they breathe out carbon dioxide. But this carbon has come from their food and that's
come from plants, which took carbon from the atmosphere. However, the carbon in
fossil fuels has remained trapped 5) _____ for hundreds of millions of years,
so it is extra carbon that's been added to the natural 6) _____. We're also
throwing away other gases into the atmosphere such as 7) _____ especially
from paddy fields and from cows, and nitrous oxide from car 8) _____.
There's extra energy trapped on the earth, already causing glaciers and ice caps to
9) _____. With more energy in the atmosphere, weather becomes
10) _____.

Energy and Future | 87

Speaking

1 Read the short passage aloud and pay attention to the idea about renewable energy.

Several countries have adopted ambitious plans to obtain their power from renewable energy (RE). These countries are not only accelerating RE installations but are also integrating RE into their existing infrastructure to reach a 100% RE mix.

Iceland gets 85% of the country's electricity from the earth's heat. The country's electricity supply is 100% renewable and depends on geothermal energy and hydropower. Norway is around 98% renewable and uses hydroelectric, geothermal and wind energy, to achieve its goal. Denmark uses 30% wind and 15% biomass for its energy needs.

In a recent study known as "The Solutions Project", Stanford professor Mark Jacobson has concluded that US can meet its 100% of energy demand through renewables by 2050 through concentrated solar power, utility-scale and rooftop photovoltaic, onshore and offshore wind, tidal and conventional hydropower and geothermal wave. Study stays to achieve its goal, conscious efforts should be made to obtain all new electricity generation by sunlight, water and wind by 2020 and US should replace 80% of its existing energy with renewable sources by 2030 to reach 100% renewable by 2050.

2 Work in pairs and answer the two questions.

1 What does Iceland's electricity supply depend on?
___
___

2 What does Stanford professor Mark Jacobson conclude in a recent study known as "The Solutions Project"?
___
___

3 Do some research on photovoltaic panels, a term you will find in the texts in this unit, and then prepare a one-minute oral presentation. Make sure the messages are properly delivered to the audience.

Extensive reading

Reading text

Gushing About America's Energy Future

1. In November, the International Energy Agency (IEA) recognized the remarkably positive developments in the US energy sector of the past few years. Not only did the IEA project that US oil production would exceed that of Saudi Arabia by 2020, it also projected that the United States would become virtually energy independent by 2035.

2. Whether or not the IEA's bold projections materialize, there can be little doubt that major change is underway in the US energy sector. At the start of the Obama administration in January 2009, the expectation was that US oil production would continue its decline of the past two decades and that oil imports would continue to increase. Instead, there has been a veritable boom in US oil and gas production that has already significantly reduced US energy import dependence.

3. Energy independence has been a foreign policy goal of successive US administrations since President Nixon set that goal in 1973 in response to a major Middle Eastern oil supply disruption that followed the Yom Kippur War. However, over the past four years, the turnaround in the US energy sector toward virtual energy independence has been driven by market-related phenomena rather than by the implementation of a national energy plan.

4. Among the more important of the market-related phenomena underlying the US energy sector's dramatic turnaround have been game-changing technological innovations including horizontal drilling and hydraulic fracturing. These innovations have allowed shale gas and "tight oil" to be extracted in a cost-efficient manner, thereby dramatically increasing US proven oil and gas reserves. In addition, major increases in fuel economy have occurred as a result of the surge in international oil prices in 2008 and the subsequent restructuring of General Motors and Chrysler following their bankruptcies.

5 The economic benefits being derived from the US move to energy independence are considerable. Over the past four years around 1.7 million oil and gas jobs have been created in the energy sector, while by 2020 one could expect three million new jobs to have been created. At the same time, there could be an appreciable narrowing in the US trade deficit as natural gas exports increase and as oil imports decline. At present, oil imports account for around two-thirds of the US trade deficit in goods. Were the United States indeed to cut oil imports by six million barrels per day by 2020, as projected by the IEA, this would save the United States around $180 billion a year at current oil prices.

6 Abundant and cheap natural gas should also support a US manufacturing revival by attracting energy-intensive industries such as aluminum, paper, iron, steel and petrochemicals to the United States. Such a revival would be supported by the fact that US natural gas prices are presently around one-third of the European level and around one-quarter of the Japanese level.

7 No less impressive would be the geopolitical gains to the United States from energy independence. No longer would the United States have to kowtow to the oil and gas producers of the Persian Gulf. OPEC's control over oil prices would be substantially diminished, while at the same time Middle Eastern power would be considerably eroded by significantly lower oil export incomes.

8 With so much to be gained both economically and politically from US energy independence, one has to hope for one thing: The US government does not now get in the way of the market-driven forces that have at last offered the United States the real prospect of energy independence.

Culture notes

horizontal drilling and hydraulic fracturing: Horizontal drilling is the practice of drilling non-vertical wells, in contrast to vertical drilling. Hydraulic fracturing is a procedure that can increase the flow of oil or gas from a well. Done together, they can make a productive well where a vertical well would have produced only a small amount of oil or gas.

shale gas and tight oil: Shale gas refers to natural gas that is trapped within shale formations. Tight oil is a type of oil found in impermeable shale and limestone rock deposits. Shales are fine-grained sedimentary rocks that can be rich sources of petroleum and natural gas. Shale gas has become an increasingly important source of natural gas in the United States since the start of the 21st century.

the Persian Gulf: The Persian Gulf, one of the most critical bodies of water, is positioned in the heart of the Middle East. The Persian Gulf and its coastal areas are the world's largest single source of crude oil.

Vocabulary

gush /gʌʃ/ v. 滔滔地说
remarkably /rɪˈmɑːkəbli/ ad. 显著地；惊人地
virtually /ˈvɜːtʃuəli/ ad. 几乎；差不多
bold /bəʊld/ a. 大胆的
underway /ˌʌndəˈweɪ/ a. 在进行中的
veritable /ˈverɪtəbl/ a. 名副其实的；十足的
successive /səkˈsesɪv/ a. 连续的；接连的；相继的
disruption /dɪsˈrʌpʃn/ n. 中断
turnaround /ˈtɜːnəˌraʊnd/ n. 好转；转机
virtual /ˈvɜːtʃuəl/ a. 实质上的；几乎的
horizontal /ˌhɒrɪˈzɒntl/ a. 水平的
hydraulic /haɪˈdrɔːlɪk/ a. 液压的；水力的
fracture /ˈfræktʃə/ v.（使）断裂；（使）折断
thereby /ˌðeəˈbaɪ/ ad. 由此；因此
surge /sɜːdʒ/ n. 陡增；激增

restructure /ˌriːˈstrʌktʃə/ v. 重新组织；调整；改组
bankruptcy /ˈbæŋkrʌptsi/ n. 破产；倒闭
considerable /kənˈsɪd(ə)rəbl/ a. 相当大的
appreciable /əˈpriːʃəbl/ a. 明显的；可觉察到的
deficit /ˈdefəsɪt/ n. 赤字
revival /rɪˈvaɪvl/ n. 复兴；复苏
energy-intensive a. 能源密集型的
petrochemical /ˌpetrəʊˈkemɪkl/ n. 石油化学产品
impressive /ɪmˈpresɪv/ a. 给人深刻印象的
kowtow /ˌkaʊˈtaʊ/ v. 顺从；惟命是从
gulf /gʌlf/ n. 海湾
diminish /dɪˈmɪnɪʃ/ v.（使）减少
erode /ɪˈrəʊd/ v. 渐渐削弱

Integrated exercises

1 Match each of the following statements with the paragraph from which the information is derived. You may choose a paragraph more than once.

_____ 1 The United States would no longer have to kowtow to the oil and gas producers of the Persian Gulf.

_____ 2 According to the projection of the IEA, US oil production would exceed that of Saudi Arabia by 2020.

_____ 3 US natural gas prices are presently around one-third of the European level and around one-quarter of the Japanese level.

_____ 4 The geopolitical benefits to the United States from energy independence are very impressive.

_____ 5 Currently oil imports make up around two-thirds of the US trade deficit in goods.

_____ 6 Energy independence has been a foreign policy goal of successive US administrations since President Nixon.

_____ 7 There has been a boom in US oil and gas production that has already greatly reduced US energy import dependence.

_____ 8 Over the past four years the energy sector in the US has helped to offer around 1.7 million oil and gas jobs.

_____ 9 Abundant and cheap natural gas should do its part toward the revival of US manufacturing by attracting energy-intensive industries.

_____ 10 Technological innovations including horizontal drilling and hydraulic fracturing have allowed shale gas and "tight oil" to be extracted in a cost-efficient manner.

2 Complete the following table to check your understanding of the major points and structure of the text.

Introduction (Paras.1-2)	There has been a veritable 1) _____ in US oil and gas production that has already significantly 2) _____ US energy import dependence.	
Body (Paras. 3-7)	Factors contributing to US energy independence	• Technological innovations have allowed shale gas and "tight oil" to be extracted in a 3) _____ manner. • Major increases in 4) _____ economy have occurred.
	Benefits derived from US energy independence	**Economic benefits:** • Millions of oil and gas 5) _____ have been created in the energy sector. • There could be an appreciable narrowing in the US 6) _____. • Abundant and cheap natural gas should also support a US manufacturing revival by attracting 7) _____ industries. **Geopolitical gains:** • The United States would no longer have to kowtow to the oil and gas producers of the Persian Gulf. • OPEC's control over oil prices would be substantially 8) _____. • 9) _____ power would be considerably eroded.
Conclusion (Para. 8)	It's hoped that the US government does not 10) _____ of the market-driven forces to realize the goal of energy independence.	

3 Work in pairs and discuss the prospective impacts of US energy independence. Take notes of the major points in your discussion while you do this task.

_____
_____
_____

Energy and Future

4 Translate the phrases into Chinese.

1. energy sector _____
2. oil production _____
3. bold projection _____
4. oil import _____
5. veritable boom _____
6. get in the way of _____
7. energy independence _____
8. oil supply disruption _____
9. dramatic turnaround _____
10. successive US administrations _____
11. horizontal drilling _____
12. hydraulic fracturing _____
13. oil and gas reserves _____
14. fuel economy _____
15. economic benefits _____
16. trade deficit _____
17. natural gas export _____
18. account for _____
19. in addition _____
20. geopolitical gains _____
21. in response to _____
22. in a cost-efficient manner _____
23. manufacturing revival _____
24. energy-intensive industry _____

5 Complete the sentences by translating the Chinese given in brackets into English. Refer to the phrases listed above when necessary.

1. This is done mainly to increase productivity, _____ (以应付国外竞争的挑战).
2. Be sure not to let your social activities _____ (妨碍你的学习).
3. The minority nationalities _____ (占人口的 6%).
4. _____ (目前的贸易逆差) indicates a serious imbalance between our import and export trade.
5. On-board microcomputers _____ (提高了燃油经济性，减少了排放量).

Intensive reading

▶ Warming up
Add subtitles to the passage to make it complete.

Solar energy works by capturing the sun's energy and turning it into electricity for your home or business.

Our sun is a natural nuclear reactor. It releases tiny packets of energy called photons, which travel the 93 million miles from the sun to the earth in about 8.5 minutes. Every hour, enough photons impact our planet to generate enough solar energy to theoretically satisfy global energy needs for an entire year.

A 2017 report from the IEA shows that solar has become the world's fastest-growing source of power – marking the first time that solar energy's growth has surpassed that of all other fuels. In the coming years, we will all be enjoying the benefits of solar-generated electricity in one way or another.

1) _____

When photons hit a solar cell, they knock electrons loose from their atoms. If conductors are attached to the positive and negative sides of a cell, it forms an electrical circuit. When electrons flow through such a circuit, they generate electricity. Multiple cells make up a solar panel, and multiple panels (modules) can be wired together to form a solar array. The more panels you can deploy, the more energy you can expect to generate.

2) _____

Photovoltaic solar panels are made up of many solar cells. Solar cells are made of silicon, like semiconductors. They are constructed with a positive layer and a negative layer, which together create an electric field, just like in a battery.

3) _____

Photovoltaic solar panels generate direct current (DC) electricity. With DC electricity, electrons flow in one direction around a circuit. For example, in a battery powering a light bulb, the electrons move from the negative side of the battery, through the lamp, and return to the positive side of the battery.

With alternating current (AC) electricity, electrons are pushed and pulled, periodically reversing direction, much like the cylinder of a car's engine. Generators create AC electricity when a coil of wire is spun next to a magnet. Many different energy sources can "turn the handle" of this generator, such as gas or diesel fuel, hydroelectricity, nuclear power, coal, wind, or solar energy.

AC electricity was chosen for the US electrical power grid, primarily because it is less expensive to transmit over long distances. However, solar panels create DC electricity. How do we get DC electricity into the AC grid? We use an inverter.

4) _____

Here's an example of how a home solar energy installation works. First, sunlight hits a solar panel on the roof. The panels convert the energy to DC current, which flows to an inverter. The inverter converts the electricity from DC to AC, which you can then use to power your home. It's beautifully simple and clean, and it's getting more efficient and affordable all the time.

However, what happens if you're not home to use the electricity your solar panels are generating every sunny day? And what happens at night when your solar system is not generating power in real time? Don't worry, you still benefit through a system called "net metering".

A typical grid-tied photovoltaic system, during peak daylight hours, frequently produces more energy than one customer needs, so that excess energy is fed back into the grid for use elsewhere. The customer gets credit for the excess energy produced, and can use that credit to draw energy from the conventional grid at night or on cloudy days. A net meter records the energy sent compared to the energy received from the grid.

▶ **Reading text**

Install Residential Solar Panels and Save

1. If you have been considering installing solar panels on your home, now is the time to do so! Sales of solar power systems have been increasing in recent years, as prices continue to fall. I would even go so far as to state that putting solar panels on your roof is like insurance for the future.

2. But should you go solar? After all, aren't solar panels expensive? Is it worth it to make the investment? The answers to all three questions are "yes". There are five reasons to install residential solar panels.

Solar electricity is free

3. Sunlight is abundant in just about every state, country and region around the world. As ultraviolet light strikes photovoltaic panels it can be converted into free solar electricity. Over the life of a solar panel system (25-35 years), you can tap

into the world's oldest renewable resource. Your investment is merely for the solar panels.

4. At current solar panel prices, taking into consideration incentives and tax credits, the retail cost of an average home solar power system in the US ranges from $20,000 to $40,000. After installation, many homeowners see their electricity bills drop by two-thirds or more – some down to zero. Plus, with net metering, homeowners with grid-tied solar power systems can watch their meters turn backward as they get credit for additional solar power generated but not used, which is fed back into the grid.

5. The return on investment for residential solar panels averages 7-10 years, which means at least 15-25 years during which you are not paying anything for solar electricity!

6. Lest you are concerned about installing solar panels because you do not live in a sunny region, consider that the leading solar power country in the world is Germany, and one of the top solar power producing states in the US is New Jersey! Solar energy can be generated wherever there is sunlight – whether filtered by clouds, or not.

Solar panels increase the value of your home

7. When most people calculate the return on investment for residential solar panels, they usually only consider the price and life of the solar power system as it depreciates over time, plus the savings in electricity bills. But another very important factor is that solar panels increase the value of your home. When you put your home on the market, they also help your house sell faster and for more money than a similarly situated property without solar panels.

8. The United States EPA reported in *The Appraisal Journal* that, for every dollar you save in annual utility costs, solar panels increase the value of your home by $20. Simple math calculations show that this can significantly add up. Let's say you save 40% of your monthly electricity bill by going solar – for us, that would be about $80 per month. Over 12 months, that is a saving of $960 which results in an impressive and immediate $19,200 boost in home equity. In other words, you instantly regain nearly 100% of the cost of a solar array in home value. A 2011 study by the National Bureau of Economic Research similarly calculates that installing residential solar panels boosts home prices by 3%-4%.

Solar power helps guard against future utility rate increases

9. Utility companies frequently raise rates and, with governmental pressure to generate increasing percentages of the power they sell from renewable resources, they seek to recoup investments in the new technology required to do so. While fossil fuel-based electricity remains cheaper than solar power or wind, government mandates cut into utility's profit margins. This means that we can expect more and greater rate hikes.

10. A solar power system can help guard against utility rate increases, however. When you generate your own electricity, you are less affected by inflation in your power bill. Even if you don't want to purchase your own solar panels, you can enter a solar panel lease or a solar power purchase agreement whereby you buy cheaper solar electricity from the owner of the panels at a fixed rate over the life of the lease or contract. These types of arrangements require no money down, as well. Once again, you can lock in existing rates and guard against future utility rate increases with solar power.

Solar power costs are reduced by incentives and tax credits

11 Currently, homeowners can save even more with residential solar panels through incentives and tax credits. In my home state of Oregon, the price of solar panels can be reduced by up to 80% as of the date of this publication. That takes an average $25,000 solar power array down to $5,600!

12 Solar power companies and other experts agree that the incentives and credits will not last forever. Several federal tax credits are set to expire at the end of 2012, if not renewed. With solar panel prices dropping at precipitous rates (over 50% reduction in the past 24 months), the purpose for continuing incentives is shrinking. So, now is probably one of the best times to invest in solar energy when you can maximize on reduced costs and generous rebates.

Going solar reduces greenhouse gas emissions

13 The vast majority of grid-based electricity comes from burning fossil fuels like coal and gas, which releases greenhouse gases into the atmosphere. That's why you can cut your carbon footprint by reducing electricity consumption. Simply, if we all used less power, demand would be reduced and utility companies would pump out fewer emissions as a result.

14 For the same reason, going solar reduces greenhouse gas emissions. Relying on renewable energy generated right at your home means you will not be as dependent – if at all – on grid-based power generated at polluting power plants. Even if you do not "believe" in global climate change, solar power is greener and safer than coal for many reasons: (1) no emissions; (2) no dangerous coal mining activities; and (3) no coal ash.

15 In summary, you can save money and the environment by going solar! Perhaps you are now convinced that going solar is the right choice for you? Consider a DIY solar panel kit, or work with a professional solar installation company, which can also help you maximize rebates and tax credits.

Culture notes

tax credit: It is an amount of money that is subtracted from taxes owed to the government. In order to encourage the use of solar energy and make solar panels affordable to more people, governments of some countries provide solar tax credits so that people can get a major discount off the price for solar panel systems.

EPA (the Environmental Protection Agency): It is an independent agency of the United States federal government whose mission is to protect human health and the environment. It has the responsibility of developing and enforcing environmental regulations under a variety of environmental laws.

Vocabulary

ultraviolet /ˌʌltrəˈvaɪələt/ a. 紫外线的
photovoltaic /ˌfəʊtəʊvɒlˈteɪɪk/ a. 光生伏打的；光电的
retail /ˈriːteɪl/ n. 零售；零卖
homeowner /ˈhəʊmˌəʊnə/ n. 房主
lest /lest/ conj. 唯恐
depreciate /dɪˈpriːʃieɪt/ v. 贬值；跌价
property /ˈprɒpəti/ n. 房产；地产
appraisal /əˈpreɪzl/ n. 评价；估价
equity /ˈekwəti/ n. 房产净值
instantly /ˈɪnstəntli/ ad. 立即；马上
regain /rɪˈɡeɪn/ v. 重新获得；复得

bureau /ˈbjʊərəʊ/ n. 局；处；署
mandate /mænˈdeɪt/ v. 命令
hike /haɪk/ n. 大幅上升
inflation /ɪnˈfleɪʃn/ n. 物价上涨；通货膨胀
lease /liːs/ n. 租约；租契
whereby /weəˈbaɪ/ ad. 由此；借此
contract /ˈkɒntrækt/ n. 合同；契约；合约
expire /ɪkˈspaɪə/ v. 到期；期满；过期
renew /rɪˈnjuː/ v. 延长…的期限；使续期
precipitous /prɪˈsɪpɪtəs/ a. 骤然的；急剧的
rebate /ˈriːbeɪt/ n. 退款
kit /kɪt/ n. 成套设备

▶ Text understanding

1 Read Paras. 1-2 and answer the following questions.

1 What is the claim made by the author at the very beginning of the text?
_____
_____

2 What are some of the doubts people have about going solar?
_____
_____
_____

3 What is the logic pattern used in the text?
_____
_____

2 Read Paras. 3-14 and complete the following table.

Five reasons to install residential solar panels	**Solar electricity is free.** • Sunlight is abundant in most parts of the world. It is the 1) _____ renewable resource. • After installation, many homeowners will find their electricity bills 2) _____ – some down to zero. • Besides, they can get credit for additional solar power generated but not used. Taken altogether, it means that at least 3) _____ you are not paying anything for solar electricity.
	Solar panels increase the value of your home. • Your house sells faster. • A study has shown that installing residential solar panels boosts home prices by 4) _____.
	Solar power helps guard against future utility rate increases. • If you generate your own electricity, you are less affected by 5) _____. • Or you can enter 6) _____ or a solar power 7) _____ whereby you buy cheaper solar electricity from the owner of the panels at a fixed rate.
	Solar power costs are reduced by incentives and tax credits. • Homeowners can save more through incentives and tax credits. • While incentives and credits are likely to 8) _____ in the near future, it may be the best time now to invest in solar energy as you can maximize on 9) _____.
	Going solar reduces greenhouse gas emissions. • Solar power is 10) _____ than coal for many reasons: no emissions; no dangerous coal mining activities; and no coal ash.

3 What methods are used to enhance the persuasiveness as well as coherence of the text? Match the paragraphs on the left with the methods on the right.

_____ Paras. 3-6   A) comparison and contrast
_____ Paras. 7-8   B) quotation
_____ Paras. 9-10   C) statistics
_____ Paras. 11-12   D) facts or examples
_____ Paras. 13-14   E) classification and division
                                          F) cause-effect

4 Read Para. 15 and answer the following questions.

1 What is the conclusion of the text?

_____

_____

2 What specific suggestion does the author offer at the end of the text?

_____

_____

▶ Language building

1 Match each of the words in the left column with its corresponding meaning in the right column.

_____	1 abundant	a to remove particular types of light
_____	2 insurance	b a continuing increase in prices, or the rate at which prices increase
_____	3 investment	c existing in large quantities; more than enough
_____	4 renewable	d a standard or level regarded as usual
_____	5 filter	e to come to an end or stop being in use
_____	6 annual	f protection against future loss
_____	7 inflation	g happening once every year
_____	8 average	h to increase sth. as much as possible
_____	9 expire	i the act of putting money or effort into sth. to make a profit or achieve a result
_____	10 maximize	j sth. that can be renewed or replaced by natural process

2 Among the three choices given, choose the one that is **NOT** close in meaning to the underlined word in each sentence.

1 There are five reasons to install residential solar panels.
   A) fix   B) set up   C) receive

2 Sales of solar power systems have been increasing in recent years, as prices continue to fall.
   A) decline   B) decrease   C) slip

3 Over 12 months, that is a saving of $960 which results in an impressive and immediate $19,200 boost in home equity.
   A) assets   B) property value C) fairness

4 Lest you are concerned about installing solar panels because you do not live in a sunny region, consider that the leading solar power country in the world is Germany.
   A) confused   B) worried   C) anxious

5 … consider that the <u>leading</u> solar power country in the world is Germany, and one of the top solar power producing states in the US is New Jersey!
   A) irrelevant   B) top   C) chief

6 When you put your home on the market, they also help your house sell faster and for more money than a similarly <u>situated</u> property without solar panels.
   A) located   B) chosen   C) placed

7 A 2011 study by the National Bureau of Economic Research similarly calculates that installing residential solar panels <u>boosts</u> home prices by 3%-4%.
   A) increases   B) promotes   C) excludes

8 Even if you don't want to purchase your own solar panels, you can enter a solar panel lease or a solar power purchase <u>agreement</u>.
   A) plan   B) deal   C) contract

9 When most people calculate the <u>return</u> on investment for residential solar panels, they usually only consider the price and life of the solar power system as it depreciates over time.
   A) profit   B) gain   C) loss

10 Over the life of a solar panel system (25-35 years), you can <u>tap into</u> the world's oldest renewable resource.
   A) utilize   B) exploit   C) turn in

3 Complete each of the following sentences with an appropriate word from the given word family. Change the form where necessary.

1 John never showed any _____ for other people's feelings. (consider, considerable, considerate, consideration)

2 We would like to _____ the loft into another bedroom. (convert, convertible, conversion)

3 The blanket will provide _____ warmth and comfort in bed. (add, addition, additional, additive)

4 Mosquitoes are extremely _____ in this dark wet place. (abound, abundant, abundance)

5 Leonard made a rapid _____: He'd never make it in time. (calculate, calculating, calculator, calculation)

6 Most of the men who gathered around him again were _____ dressed. (similar, similarly, similarity)

7 It will produce _____ more cheaply than a nuclear plant. (electric, electron, electrical, electricity)

8 She needs to _____ a sense of her own worth. (gained, gainful, regain)

9 Cells divide and _____ as part of the human growth process. (new, renew, renewal, renewable)

10 He was advised to reduce his alcohol _____. (consume, consumer, consumable, consumption)

4 Match each word in the box with the group of phrases where it is usually found.

| solar | strike | resource | range | result | utility |

1 _____
 renewable ~s
 scarce ~s
 natural ~s
 financial ~s

2 _____
 ~ from … to …
 ~ in size / length / price
 ~ between … and …
 ~ widely

3 _____
 light ~s photovoltaic panels
 light ~s the face
 sun ~s the glass
 sun ~s the fields

4 _____
 ~ panels
 ~ power
 ~ heating
 go ~

5 _____
 ~ costs
 ~ rate
 ~ company
 ~ bill

6 _____
 ~ in
 as a ~
 ~ from
 ~ of

5 Find the idiomatic expressions in the text matching the Chinese equivalents.

1 零售成本 _____
2 住宅太阳能板 _____
3 转换成 _____
4 可再生资源 _____
5 电费账单 _____
6 计算回报 _____
7 绝大多数 _____
8 提高收费标准 _____
9 考虑到 _____
10 利用；开发 _____
11 专业安装公司 _____
12 急剧下降 _____
13 电网供电 _____
14 削减利润率 _____
15 税收抵免 _____
16 经…过滤 _____
17 逐渐贬值 _____
18 类似地段的地产 _____
19 以固定的费率 _____
20 减少碳足迹 _____

6 Combine the sentences given below. Then compare your sentences with the original ones in the text.

1 _____

 a At current solar panel prices, the retail cost of an average home solar power system in the US ranges from $20,000 to $40,000.
 b Incentives and tax credits are taken into consideration in the retail cost.

Energy and Future | 103

2 _____

   a With net metering, homeowners with grid-tied solar power systems can watch their meters turn backward as they get credit for additional solar power generated but not used.

   b The additional solar power is fed back into the grid.

3 _____

   a You are concerned about installing solar panels because you do not live in a sunny region.

   b If you are concerned, consider two facts.

   c One fact is that the leading solar power country in the world is Germany.

   d The other fact is that one of the top solar power producing states in the US is New Jersey!

4 _____

   a Even if you don't want to purchase your own solar panels, you can enter a solar panel lease or a solar power purchase agreement.

   b By means of the lease or purchase agreement, you buy cheaper solar electricity from the owner of the panels at a fixed rate over the life of the lease or contract.

5 _____

   a The vast majority of grid-based electricity comes from burning fossil fuels like coal and gas.

   b Burning fossil fuels releases greenhouse gases into the atmosphere.

7 Translate each of the Chinese sentences into English by using the underlined phrase or structure in the example.

1 I would even <u>go so far as to</u> state that putting solar panels on your roof is like insurance for the future.
我才不会说他是个骗子。

_____

_____

2 <u>Taking into consideration</u> incentives and tax credits, the retail cost of an average home solar power system in the US ranges from $20,000 to $40,000.
时间因素是我们必须首先考虑的。

_____

_____

3 <u>Lest</u> you are concerned about installing solar panels because you do not live in a sunny region, consider that the leading solar power country in the world is Germany.
让我先记下你的电话号码，以免忘了。
_____
_____

4 When you <u>put</u> your home <u>on the market</u>, they also help your house sell faster.
新型彩电不久将投放市场。
_____
_____

5 A solar power system can help <u>guard against</u> utility rate increases.
安全摄像头已被安装用来防范偷猎者。
_____
_____

6 With solar panel prices dropping <u>at</u> precipitous <u>rates</u>, the purpose for continuing incentives is shrinking.
报警案件的数量正在以惊人的速度增长。
_____
_____

Academic writing

▶ Micro-skill: References and quotations

Academic writing usually begins with the research and ideas of others, so it is necessary to show which sources you have cited in your work.

1 References

A reference is an acknowledgement that you have read other writers' ideas or data for your reference. They are either in the form of a summary / paraphrase or a quotation. This way, links are provided in the reference list at the end of the paper. For example:

Of all the non-verbal forms of communication, Leatrice Eiseman (2000) notes, color is the most immediate means of communicating messages and meanings. J. H. Kleynhans observes that "color stimulates and works synergistically with all of the senses, symbolizes abstract concepts and thoughts, expresses fantasy or wish fulfillment, recalls another time or place and produces an aesthetic or emotional response". (J. H. Kleynhans, 2007: 46)

List of references
Eiseman, L. (2000). *Pantone Guide to Communicating with Color.* Sarasota: Grafix Press Ltd.
Kleynhans, J. H. (2007). The Use of Color as a Tool for Propaganda. *Interim: Interdisciplinary Journal*, 6(1), 46-53

In the example above, which is a summary and which is a quotation?

A quotation should include the writer's name, year of publication and page number, e.g., (Smith, 2009: 37), while a summary is made up of the writer's name and year of publication, e.g., Smith (2009).

Generally, summaries and quotations are introduced by reference verbs. For example:

Melville (2009) assumes that …
Johnson (1972) observed that …

Reference verbs can be used either in the present or the past tense. The present tense indicates that the source is recent and still valid, while the past tense suggests that the source is older and may be outdated, but there are no clear distinctions. In some disciplines, an old source is still valid.

The following Latin abbreviations are sometimes used in references.

et al. (used when there are three or more writers and the full list of names is given in the reference list)	Fossil fuels will eventually be replaced by renewable energy (Smith et al., 2006: 137).
ibid. (taken from the same source, i.e., the same page, as the previous citation)	Fossil fuels are more likely to cause severe pollution (ibid.) …
op. cit. (taken from the same source as previously, but a different page)	Fletcher (op. cit.) has commented that
n.d. (for sources that do not have a date of publication, and often used after the writer's name)	Southey, R. (n.d.). *The life of Nelson*. London: Blackie & Son Ltd.

2 Quotations

Using quotations indicates that you can include the original words of a writer in academic writing. However, quotations must not be overused. They can be used when the original ideas are distinctive or there is no better way to deliver them or when the original words are well-known and do not need any explanation.

Quotations are often used together with phrases that show the sources, and explain how a quotation is appropriate for the argument. For example:

Introductory phrase	Writer	Reference verb	Quotation	Reference
Many people agree to it;	just like J. H. Kleynhans	observes that	"color stimulates and works synergistically with all of the senses, symbolizes abstract concepts and thoughts, expresses fantasy or wish fulfillment, recalls another time or place and produces an aesthetic or emotional response".	(2007: 46)

Tip 1: Quotation marks are used for short quotations (2-3 lines). For example:

As Richardson (2007: 21) summarizes, "there are different – and sometimes radically different – accounts of what discourse is," resulting in a range of overlapping and sometimes contrasting theorizations (Fairclough, 2003).
Glaser (1999) takes some of these debates outside the traditional urban context, pointing out that "rural communities in America maintained their sense of identity

and place by accessing electricity in ways that allowed them to integrate these changes on their own terms" (p. 12).

Tip 2: As for longer quotations, they are either indented (given a wider margin) or are printed in smaller type. No quotation marks are needed. For example:

Historian George Lipsitz (2012: 33) argues that technological, social, and economic changes dating from the nineteenth century have culminated in what he calls a "crisis over the loss of connection to the past", in which Americans find themselves cut off from the memories of their traditions.

> The transformations in behavior and collective memory fueled by the contradictions of the nineteenth century have passed through three major stages in the United States. The first involved the establishment and codification of commercialized leisure from the invention of the telegraph to the 1890s. The second involved the transition from Victorian to consumer-hedonist values between 1890 and 1945. The third and most important stage, from World War II to the present, involved extraordinary expansion in both the distribution of consumer purchasing power and in the reach and scope of electronic mass media. The dislocations of urban renewal, suburbanization, and deindustrialization accelerated the demise of tradition in America, while the worldwide pace of change undermined stability elsewhere. The period from World War II to the present marks the final triumph of commercialized leisure, and with it an augmented crisis over the loss of connection to the past.

Tip 3: After the year of publication, there should be page numbers, for example, (1974: 93)

Tip 4: Quotations should be exactly the original words of the writer. When it is necessary to delete some words that are irrelevant, use ellipsis to replace the unwanted words.

> "The Internet … has been regarded as one of the greatest inventions of the century."

Tip 5: Square brackets [ ] can be used to add or change words in the quotation to clarify a point.

> "It [driving] imposes a heavy procedural workload on cognition that … leaves little processing capacity available for other tasks."

Study the following passage from an article titled "How the energy transition will reshape geopolitics" in the journal *Nature* (Vol. 569, p. 29-31, 2019) by Goldthau et al. p. 29.

Energy is at the root of many political ructions. President Donald Trump's intention to pull the United States out of the Paris climate agreement in 2020, the European Union's restrictive policies against importing photovoltaic cells and the political hostility towards the school strikes over climate-change inaction are all reactions to attempts to shift the world to a low-carbon economy. The future benefits of clean energy can seem distant when weighed against pay packets or votes now. Despite the impacts of climate change becoming increasingly evident in devastating cyclones, heatwaves and floods, politicians want to protect local jobs and incumbent industries, such as coal and manufacturing.

1 Write a summary of the paragraph, including a reference.

_____
_____
_____

2 Write a summary of the paragraph with a quotation.

_____
_____
_____

3 Write a summary of the paragraph with both a reference and a quotation.

_____
_____
_____

Macro-skill: Cause-effect

Cause-effect analysis is another common way of developing an essay. Causes of an event, fact or phenomenon, fall into various types. In view of the relevance of the cause to the effect, some causes are *immediate* or *direct* ones while others can be *indirect* or *ultimate*. *Immediate / Direct causes* can be identified without much effort and are directly associated with the effect they produce. *Indirect / Ultimate causes*, on the contrary, are more difficult to uncover, but they are the basic, underlying factors that help to explain the results.

The cause-effect pattern may follow several types based on their correspondence, i.e., one cause – one effect, one cause – multiple effects, multiple causes – one effect, and causal chain, as presented below.

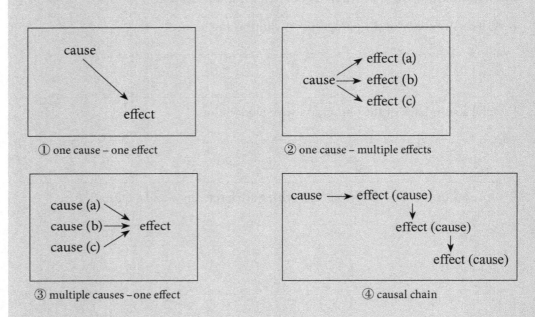

Two points are to be noted here.

1 The cause-effect relationship is a term for logical reasoning. In academic writing, the cause can go either before or after the effect. In other words, the effect / result can appear either at the beginning of an essay or at its end. In the former case, the effect / result serves as the topic of an essay while in the latter, as its conclusion.

2 Causal analysis often serves as an effective tool for debating over or justifying an idea in an essay. The debate or justification may sound strong if the causes are derived from facts.

1 Read the following passage and complete the table.

Americans are proud of their variety and individuality, yet they love and respect few things more than a uniform, whether it is the uniform of an elevator operator or the uniform of a five-star general. Why are uniforms so popular in the United States?

Among the arguments for uniforms, one of the first is that in the eyes of most people they look more professional than civilian clothes. People have become conditioned to expect superior quality from a person who wears a uniform. The television repairman who wears a uniform tends to inspire more trust than one who appears in civilian clothes. Faith in the skill of a garage mechanic is increased by a uniform. What easier way is there for a nurse, a policeman, a barber, or a waiter to lose professional identity than to step out of uniform?

Uniforms also have many practical benefits. They save on other clothes. They save on laundry bills. They are tax-deductible. They are often more comfortable and more durable than civilian clothes.

Effect (Topic)	1) _____.
Causes	• 2) _____. Examples: 3) _____. 4) _____. • 5) _____. Examples 6) _____. 7) _____. 8) _____.

2 Read the following passage and complete the table.

Instead of retiring to Florida, seniors are increasingly choosing to spend their golden years back in the classroom. These lifelong learners are boxing up their possessions and moving in large numbers to college towns across the country, where they enjoy easy access to many cultural, educational and recreational opportunities available on campus.

College towns are becoming more popular than they were 10 or 20 years ago. They offer a wide range of convenience, typically without air pollution, traffic jams, and high crime rates often found in larger cities. And the cost of living is much lower in smaller college cities.

Effect (Topic)	1) _____.
Causes	• 2) _____.
	• 3) _____.
	• 4) _____.
	• 5) _____.

3 Read the following passage and complete the table.

Each year over 600 million people travel internationally. Hundreds of millions more journeys are within their home country, doing so for both work and pleasure. As a result, the tourist industry – including hotels, scenic spots, airlines, travel agencies and other businesses – is described as "the world's number one employer".

What are the origins of this booming industry? The seeds of the modern tourist industry in the West were sown especially in the 19th century. As the industrial revolution swelled the ranks of the middle classes in Europe and the United States, a growing number of people found themselves with both the money and the time to travel. In addition, great advances were made in methods of mass transportation. Powerful engines pulled passengers between major cities, and great steamships sped them between continents. To deal with the growing traffic, large hotels sprang up near railway terminals and shipping ports.

Effect (Topic)	1) _____.
Causes (Origins)	• 2) _____.
	• 3) _____.
	Examples:
	4) _____.
	5) _____.
	• 6) _____.

Writing assignment

Choose one of the topics below and compose a short essay of about 120 words. Use the pattern of cause-effect.

1 Would you suggest installing solar energy systems to your family? Give specific reasons to explain.

2 Do you agree that renewable energy is the future trend of energy use? Support your viewpoint with persuasive reasons.

Sharing

You are going to hold a debate on your prediction of the US energy future. The standpoints of the two sides are as follows.
Side A: The US will become energy independent due to shale gas.
Side B: The US will mainly rely on solar and other renewables.

1. Form a group of four and debate with neighboring groups holding opposing ideas.

2. List your arguments and your opponents' potential arguments for the debate. You also have to collect evidence to support your standpoints.

3. Run the debate according to the following procedure.

1) Present your standpoints to the audience;
2) Take turns elaborating on your arguments based on your evidence and proof and challenge your opponents;
3) Make a summary of your statements.

4. After the debate, vote for the winning team and the best debater in each debate.

Unit 5 Energy and Transformation

Lead-in

A shift from conventional to non-conventional energy resources such as renewables and shale gas takes time and is far from easy. Appealing as it appears, wide application of wind and solar power is deemed unrealistic by many in that it has inherent problems that are hard to solve. What are these problems?

With the development of horizontal drilling and hydraulic fracturing, considerable reserves of shale gas, as a form of fossil fuels different from the conventional gas, have been discovered around the world. Following the example of the US, Australia hopes to exploit this resource on a large scale, but it seems no easier than developing renewables. What obstacles are Australians faced with?

Learning objectives

Upon completion of this unit, you will be able to:
- talk about major problems of developing wind and solar power;
- analyze the potential for a shale revolution in Australia;
- use generalizations properly in academic writing;
- recognize and use the pattern of classification in writing;
- make an analysis report on China's potential to develop renewables and shale gas.

Viewing and speaking

▶ Viewing

> **New words**
>
> hydroelectric /ˌhaɪdrəʊˈɪlektrɪk/ a. 水力发电的
> dilute /ˈdaɪluːt/ a. 稀释的
> intermittent /ˌɪntəˈmɪtnt/ a. 间歇的
> phosphorus /ˈfɒsfərəs/ n. 磷
>
> boron /ˈbɔːrɒn/ n. 硼
> titanium /taɪˈteɪniəm/ n. 钛
> neodymium /ˌniːəʊˈdɪmɪəm/ n. 钕
> erratic /ɪˈrætɪk/ a. 不稳定的

1 Watch Part I of the video clip and decide whether the statements are true (T) or false (F).

☐ 1 The video aims to answer why wind and solar power only constitute a very small part of the world's energy source.

☐ 2 For an energy resource to be popularized, it has only to be cheap.

☐ 3 Although sunlight and wind are free, it costs a lot to turn them into usable energy in a large quantity.

☐ 4 Wind and solar power have two basic problems: diluteness and intermittency.

☐ 5 Diluteness means sunlight and wind are not constant, and intermittency means they are weak.

2 Watch Part II of the video clip and fill in the blanks with what you hear.

To solve the diluteness problem, we need a lot of additional materials to produce a unit of energy. For solar power, such materials can include highly 1) _____ silicon, phosphorus, boron and a dozen other complex compounds like titanium dioxide. For wind, needed materials include high-performance compounds and the rare earth metal neodymium as well as steel and 2) _____. And as big a problem as diluteness is, it's nothing compared to the intermittency problem. The only way for solar and wind to be truly useful is to 3) _____ them. There is not one real or proposed, 4) _____, free-standing solar or wind power plant. Wind constantly varies, sometimes 5) _____ completely. And solar produces little in the 6) _____ months when Germany most needs energy, and therefore some reliable source of energy is needed. While Germany has spent tens of billions of dollars on solar panels and windmills, fossil fuel used in the nation has not 7) _____; it's increased. And less than 10% of Germany's total energy is 8) _____ by solar and wind. Furthermore, 9) _____ back and forth between solar and wind and coal to maintain 10) _____ energy is costly. There is no free lunch and there is no free energy. That very much includes the highly expensive energy from the sun and the wind.

Speaking

1 Read the short passage aloud and pay attention to the idea about solar trees.

Indian scientists have designed a new device that they hope will solve one of the biggest problems with the use of solar energy. They call the device a solar tree. Solar trees have metal "branches" extending from a tall, central pole at different levels. Each branch holds a photovoltaic panel, also called a solar panel. These panels are flat surfaces that collect energy from the sun and turn it into electricity.

The shape of the solar trees makes it possible to fit more photovoltaic panels in a space than traditional systems do. This means less land would be needed to produce solar energy. The space-saving tree will make it easier to provide solar energy to homes in cities. The trees will also take less space from farmers in rural areas. Solar trees will collect more energy than normal photovoltaic panels on the tops of buildings, too. The device facilitates the placement of solar panels in a way that they are exposed more toward the sun. So they are able to harness 10%-15% more energy.

2 Work in pairs and answer the two questions.

1   What are the advantages of solar trees?

_____

_____

2   What is the function of solar panels?

_____

_____

3 Do some research on wind turbines, a term you will find in the texts in this unit, and then prepare a one-minute oral presentation. Make sure the messages are properly delivered to the audience.

Extensive reading

Reading text

Wind and Solar Power: Uncooperative Reality

1. Energy policies in the United States for decades have pursued energy sources defined in various ways as alternative, unconventional, independent, renewable, and clean in an effort to replace such conventional fuels as oil, coal and natural gas. These long-standing efforts have, without exception, yielded poor outcomes, in a nutshell because they must swim against the tide of market forces. That is why the only reliable outcome has been one disappointment after another, and there are powerful reasons to predict that the same will prove true with respect to the current enthusiasm for renewable electricity. In this article, I will look at the inherent limitations of wind and solar power generation – that is, the reasons why they are likely to remain uncompetitive without very large subsidies and other policy support, or even with such support.

Unconcentrated energy content

2. The energy content of wind flows and sunlight, which varies depending upon air speed and sunlight intensity, is far less concentrated than that of the energy contained in fossil or nuclear fuels. To compensate for this unconcentrated nature of renewable energy sources, the facility operator or power utility must invest large amounts in land and materials to make renewable generation technically practical for generating nontrivial amounts of electricity. A wind farm would require 500 windmills of two megawatts (MW) each to provide a theoretical generation capacity of 1,000 MW. Because the wind turbines must be spaced apart to avoid wake effects (wind interference among the turbines), a 1,000 MW wind farm would require on the order of 48,000-64,000 acres of land. With an assumed capacity factor for a typical wind farm of 35%, reliable wind capacity of 1,000 MW would require an amount of land on the order of three times that rough estimate. In contrast, a 1,000 MW gas-fired plant requires about 10-15 acres; conventional coal, natural gas and nuclear plants have capacity factors of 85%-90%.

3. The same general problem afflicts solar power. The energy content of sunlight, crudely, is about 150-400 watts per square meter depending on location, of which about 20%-30% is convertible to electricity depending on the particular technology. Accordingly, even in theory, a square meter of solar energy-receiving capacity is enough to power roughly one 100-watt light bulb, putting aside such issues as sunlight intensity. This problem of land requirements for solar thermal facilities is of sufficient importance that most analyses assume a maximum generation capacity of 50-100 MW, which, conservatively, would require approximately 1,250 acres, or two square miles.

4. In short, transformation of the unconcentrated energy content of wind and sunlight into a form usable for modern applications requires massive capital investment in land and wind turbines and solar-receiving equipment. This means that energy from renewable sources, relative to that from conventional forms, by its very nature is limited and expensive.

Siting limitations and transmission costs

5. Conventional power-generation plants can be sited, in principle, almost anywhere, and such fuels as coal and natural gas can be transported to the generation facilities. This means that investment-planning decisions can optimize transmission investment costs along with the other numerous factors that constrain and shape generation investment choices: land costs, environmental factors, reliability issues, transmission line losses, and the like. Wind and solar sites, on the other hand, must be placed where the wind blows and the sun shines with sufficient intensity and duration. Because appropriate sites are limited, with the most useful ones exploited first, the cost of exploiting such sites must rise as more sites are used. As a result, even if wind and solar technologies exhibit important scale economies in terms of capacity and / or generation costs, scale economies may not be present at the industry level as developers are forced onto increasingly unfavorable sites.

6. Because conventional power-generation investments can optimize transmission costs and other reliability factors more easily than is the case for wind and solar capacity, it would be surprising if such costs were not higher for the latter. This is exacerbated by the physical realities that wind conditions are strongest in open plains regions, while solar generation generally requires regions with strong sunlight and, for thermal solar plants, sizable open areas. For the United States, the best wind capacity sites are in a region stretching from the northern plains down through Texas, and the best thermal solar sites are in the Southwest. The United States simply lacks significant east-west high-voltage interconnection transmission capacity to transport wind and solar power to the coasts.

Low availability and high cost

7. Electric supply systems respond to growing demands ("load") over the course of a day or year by increasing output from the lowest-cost generating units first and then calling upon successively more expensive units as electric loads grow toward the daily (or seasonal) peak. Because of the uncertainties caused by the unreliability of wind and sunlight, most electric generation capacity fueled by renewable energy sources cannot be assumed to be available on demand when their use is expected to be most economic. Accordingly, this capacity cannot be scheduled (or "dispatched"). Instead, it requires backup generation capacity to preserve system reliability.

8. Cost estimates published by the Energy Information Administration (EIA) suggest that this backup capacity has imposed total costs per mWh of about $368. The estimated wind and solar costs in 2016 are about $149 and $257-$396 per mWh, respectively; if we add the rough estimate for backup costs, the total is about $517 for wind and $625-$764 for solar generation. The EIA estimates for gas- or coal-fired generation are about $80-$110 per mWh. Accordingly, the projected cost of renewable power in 2016, including the cost of backup capacity, is at least five times higher than that for conventional electricity.

9. The higher cost of electricity generated with renewable energy sources is only one side of the competitiveness question; the other is the value of that generation, as not all electricity is created equal. In particular, power produced at periods of peak demand is more valuable than off-peak generation. In this context, wind generation, in particular, is problematic because, in general, winds in the US tend to blow at night and in the winter, which corresponds inversely to peak energy demand during daylight hours and in the summer.

Conclusion

10. The most striking economic characteristic of wind and solar power is a small market share stubbornly resistant to strong political support and very substantial direct and indirect subsidies. This general problem of poor competitiveness is the result of factors intrinsic to renewable electricity: unconcentrated energy content, siting limitations and additional resulting transmission costs, and poor reliability and the attendant costs of backup capacity. These problems are inherent in the technologies and can be overcome only at considerable expense.

Culture notes

wake effect: It is the aggregated influence on the energy production of the wind farm, which results from the changes in wind speed caused by the impact of the turbines on each other. It is important to consider wake effects from neighboring wind farms and the possible impact of wind farms which will be built in the future.

Energy Information Administration (EIA): It is the statistical agency of the US Department of Energy. It collects, analyzes and disseminates independent and impartial energy information to promote sound policymaking, efficient markets, and public understanding of energy and its interaction with the economy and the environment.

Vocabulary

uncooperative /ˌʌnkəʊˈɒp(ə)rətɪv/ *a.* 不合作的；不配合的
unconventional /ˌʌnkənˈvenʃn(ə)l/ *a.* 非常规的
yield /jiːld/ *v.* 产生；得出
in a nutshell 简而言之
enthusiasm /ɪnˈθjuːziˌæzəm/ *n.* 热忱；热情
inherent /ɪnˈherənt; ɪnˈhɪərənt/ *a.* 固有的；内在的
limitation /ˌlɪmɪˈteɪʃn/ *n.* 局限；弱点
uncompetitive /ˌʌnkəmˈpetətɪv/ *a.* 无竞争力的
subsidy /ˈsʌbsədi/ *n.* 补贴；津贴；补助金
unconcentrated /ʌnˈkɒnsnˌtreɪtɪd/ *a.* 不集中的
intensity /ɪnˈtensɪti/ *n.* 强度
technically /ˈteknɪkli/ *ad.* 技术上
nontrivial /nɒnˈtrɪvɪəl/ *a.* 有用的
megawatt /ˈmegəˌwɒt/ *n.* 兆瓦；百万瓦特
interference /ˌɪntəˈfɪərəns/ *n.* 干扰
afflict /əˈflɪkt/ *v.* 折磨；使苦恼
crudely /ˈkruːdli/ *ad.* 天然地；未经提炼地
watt /wɒt/ *n.* 瓦特

convertible /kənˈvɜːtəbl/ *a.* 可转换的；可转变的
accordingly /əˈkɔːdɪŋli/ *ad.* 因此；所以
conservatively /kənˈsɜːvətɪvli/ *ad.* 保守地
constrain /kənˈstreɪn/ *v.* 限制；约束
duration /djʊˈreɪʃn/ *n.* 持续时间
exploit /ɪkˈsplɔɪt/ *v.* 开发；利用
exhibit /ɪgˈzɪbɪt/ *v.* 显示；展现
unfavorable /ʌnˈfeɪv(ə)rəbl/ *a.* 不适宜的
stretch /stretʃ/ *v.* 延伸；绵延
seasonal /ˈsiːz(ə)nəl/ *a.* 季节性的
unreliability /ˈʌnrɪˌlaɪəˈbɪləti/ *n.* 不可靠性
respectively /rɪˈspektɪvli/ *ad.* 分别地；各自地
competitiveness /kəmˈpetɪtɪvnɪs/ *n.* 竞争力
off-peak /ˈɒfˌpiːk/ *a.* 非高峰期的
problematic /ˌprɒbləˈmætɪk/ *a.* 成问题的
inversely /ˌɪnˈvɜːsli/ *ad.* 相反地；倒转地
striking /ˈstraɪkɪŋ/ *a.* 显著的
resistant /rɪˈzɪst(ə)nt/ *a.* 不受影响的
substantial /səbˈstænʃl/ *a.* 大量的
intrinsic /ɪnˈtrɪnsɪk/ *a.* 本质的；内在的
attendant /əˈtendənt/ *a.* 伴随的；随之而来的

Integrated exercises

1 Match each of the following statements with the paragraph from which the information is derived. You may choose a paragraph more than once.

_____ 1 Land requirement is a problem for solar energy utility.

_____ 2 Transmission for solar-generated and wind-generated electricity costs more money than conventional power transmission.

_____ 3 The projected cost of renewable power could be five times higher than that for conventional electricity in 2016.

_____ 4 Energy from renewable sources by its very nature is limited and expensive.

_____ 5 The most striking feature of wind and solar power is a small market share versus strong political and economic support.

_____ 6 The United States lacks east-west high-voltage interconnection transmission capacity.

_____ 7 Unreliability of wind and sunlight makes backup generation capacity necessary to meet peak demand.

_____ 8 US energy policies are against the market forces.

_____ 9 The facility operator or power utility must invest large amounts of money to make renewable generation technically practical.

_____ 10 Generally speaking, periods of peak wind generation doesn't match periods of peak energy demand.

2 Complete the following table to check your understanding of the major points and structure of the text.

Introduction (Para. 1)	Wind and solar power generation are likely to remain 1) _____.
Body (Paras. 2-9)	**Unconcentrated energy content** Compared with fossil or nuclear fuels, wind flows and sunlight are far less 2) _____, so the facility operator or power utility must 3) _____ large amounts in land and materials to make renewable generation technically practical. **Siting limitations and transmission costs** • Wind and solar sites, must be 4) _____ where the wind blows and the sun shines with 5) _____ intensity and duration. It is difficult for wind and solar power generation investments to 6) _____ transmission costs and other reliability factors.

(to be continued)

(continued)	
Body (Paras. 2-9)	**Low availability and high cost** • Because wind and sunlight are unreliable, its electric generation capacity is not 7) _____ on demand when electric loads grow and economic electricity is expected. • To preserve system reliability, renewable power requires 8) _____ , and costs at least 9) _____ higher than conventional electricity. • Power produced at periods of peak demand is 10) _____ than off-peak generation. Wind generation is problematic in this context.
Conclusion (Para. 10)	The problem of poor competitiveness of wind and solar power results from factors 11) _____ to renewable electricity: unconcentrated energy content, siting limitations and additional transmission costs, and poor reliability and the attendant costs of backup capacity.

3 Complete the following table with the differences between conventional electricity and renewable electricity.

	Conventional electricity	Renewable electricity
Energy resources	Fossil or nuclear fuels	Wind flows and sunlight
Land requirements	A 1,000 MW gas-fired plant requires about 1) _____ .	Reliable wind capacity of 1,000 MW would require an amount of land on the order of 2) _____ times that rough estimate.
Capacity factors	Conventional coal, natural gas and nuclear plants have capacity factors of 3) _____ .	An assumed capacity factor for a typical wind farm is about 4) _____ .
Siting limitations	Conventional power-generation plants can be sited almost 5) _____ .	Renewable power plants usually need 6) _____ open areas.
Availability	No need for backup generation capacity	Due to 7) _____ , backup generation capacity is needed.
Cost		Higher, with additional resulting transmission costs and 8) _____ of backup capacity

4 Work in pairs and discuss your ideas about the limitations of wind and solar power generation. Take notes of the major points in your discussion while you do this task.

_____
_____
_____

5 Translate the phrases into Chinese.

1 long-standing effort _____
2 in a nutshell _____
3 power utility _____
4 swim against the tide _____
5 facility operator _____
6 wake effect _____
7 capacity factor _____
8 on the order of _____
9 in contrast _____
10 transmission line _____
11 scale economy _____
12 physical reality _____
13 call upon _____
14 put aside _____
15 available on demand _____
16 backup generation capacity _____
17 system reliability _____
18 peak demand _____
19 off-peak generation _____
20 correspond inversely to _____
21 resistant to _____
22 indirect subsidy _____
23 attendant cost _____
24 the only reliable outcome _____

6 Complete the sentences by translating the Chinese given in brackets into English. Refer to the phrases listed above when necessary.

1 Our competitiveness and market share benefit from _____ (我们长期的努力).
2 This was the most effective way to deploy our resources, particularly _____ (在需求峰值期间).
3 These factories _____ (实现了规模经济) through mass production.
4 He was _____ (不怕与大众观点不一致).
5 These crops contain more vitamins and will be more _____ (不受害虫侵害).

Energy and Transformation | 123

Intensive reading

▶ Warming up

1 We talk a lot about the shale gas. What is shale gas? Why is it so significant for the energy industry? The following passage may help clarify these questions. Choose suitable words from the box to fill in the blanks. Change the form where necessary.

| access | gas | rock |
| produce | possess | increase |

Shale gas refers to natural 1) _____ that is trapped within shale formations. Shales are fine-grained sedimentary 2) _____ that can be rich sources of petroleum and natural gas. Over the past decade, the combination of horizontal drilling and hydraulic fracturing has allowed 3) _____ to large volumes of shale gas that were previously uneconomical to 4) _____. The production of natural gas from shale formations has rejuvenated the natural gas industry in the United States. According to the EIA *Annual Energy Outlook 2011*, the United States 5) _____ 2,552 trillion cubic feet (Tcf) of potential natural gas resources. At the 2009 rate of US consumption (about 22.8 Tcf per year), 2,552 Tcf of natural gas is enough to supply approximately 110 years of use. Shale gas resource and production estimates increased significantly between the 2010 and 2011 *Outlook* reports and are likely to 6) _____ further in the future.

2 Work in groups and discuss the challenges of shale gas exploration and where shale gas comes from.

▶ Reading text

1 Australia, home to the world's seventh largest recoverable shale gas reserves, has several characteristics conducive for commercializing the resource including existing infrastructure, industry know-how and low population density in shale-rich regions.

2 So, what is the potential for a US-like shale revolution in the country?

3 Australia has an estimated 437 Tcf of recoverable shale gas reserves, according to the EIA, around two-thirds of the US' 665 Tcf – the world's fourth largest. "Broadly most people would recognize that Australia is the closest analogy to the US, with our infrastructure position and unconventional shale gas opportunities," James Baulderstone, Vice President Eastern Australia at oil and gas firm Santos, said.

4 The Australian shale gas industry is still in its infancy, but exploration has increased in the last few years. The sector has been drawing international interest from global players. The likes of Chevron, ConocoPhillips, Statoil, Total, BG Group, have invested over $1.55 billion in Australia's shale gas industry as of mid-2013, according to the EIA.

Is a US-style Shale Revolution Coming for Australia?

5 Santos is among a group of Australian companies developing shale properties in the country. It launched the country's first commercial shale gas well near its conventional gas activities in the Cooper Basin in South Australia, in 2012. There is currently limited commercial shale gas production in Australia, however.

6 Industry analysts say while Australia's shale sector holds promise, its challenges are very different to the US in terms of production costs, the onshore service industry and the regulatory environment – which could constrain the industry's potential. "I think it's probably harder in Australia than it is in the US for a number of reasons," said Alex Wonhas, Executive Director Energy and Resources at Commonwealth Scientific and Industrial Research Organisation (CSIRO).

7 For starters, developing shale resources in Australia is costlier than in the US, Wonhas noted. In Australia, several basins are located in remote parts of the country. While this may mean less environmental opposition, these areas would require significant investments in pipeline and processing infrastructure.

8 While the infrastructure in the Cooper Basin is reasonably good, there are no pipelines connecting the Canning, Georgina and Officer basins to the existing main transmission lines, according to the University College London (UCL) International Energy Policy Institute. Pipelines are seen as the most effective means of transporting gas directly to either consumers or a liquefied natural gas (LNG) terminal.

9 Furthermore, the capacity of Australia's service industry in terms of drilling and hydraulic fracturing services is not adequate, say experts. "In Australia, there's limited service sector availability, in terms of fracking crews and high specification rigs that can drill deep into the shales," said Chris Graham, head of Australasia upstream research at Wood Mackenzie. Shale gas drilling and fracking are over three times as costly as in the US, according to industry estimates.

10 Another key difference is the regulatory environment, in particular the mineral rights ownership, say industry participants. In the US, landowners possess the rights to the resources beneath their land and are entitled to royalties. This has ensured local communities are able to benefit financially, thus helping to temper local opposition to the industry.

11 In Australia, the state owns any underground

resources. Australian landholders have to provide access, in return for some compensation, to energy companies that want to explore and exploit their land. The proceeds are generally not as large as US royalty payments, according to RFC Ambrian, a firm providing corporate financial advisory services in the natural resources market, so there is less incentive for landholders to encourage minerals development on their properties.

12 Factors such as mineral rights ownership may slow down the development of the Australia's shale deposits, said Pal Haremo, Vice President Exploration at Norwegian oil and gas company Statoil, which has invested in shale acreage in the country with Canada's PetroFrontier. "It's not just about the quality of the rock; it's also about commercial terms, politics. The US has a unique system where landowners own the minerals, so there's a win-win."

13 Finally, there remains a great deal of uncertainty around the quality of Australia's shale resource as reserves haven't been tapped in a big way yet, making investing in the sector a high risk proposition.

14 "Just because you've understood a play in the US, it doesn't mean you can necessarily translate that into an Australian play. You can only understand it by drilling it and producing it to see what the drop off in production is. And that's very location and shale specific," said Graham of Wood Mackenzie.

15 This year, some clarity will begin to emerge as a number of companies will publish flow test results from their shale wells, which will provide an indication of their production potential. As such, Graham says it may be too early to draw conclusions about the industry's potential. "We're so early in the day. There have been some positive results in some areas, but slow progress in others. It's really too early to tell," he said.

Culture notes

Santos (South Australia Northern Territory Oil Search): It is one of the leading independent oil and gas producers in the Asia-Pacific region, supplying energy to homes, businesses and major industries across Australia and Asia.

Cooper Basin: It is located mainly in the southwestern part of Queensland and extends into northeastern South Australia. It is the most important onshore petroleum and natural gas deposits in Australia.

Vocabulary

conducive /kənˈdjuːsɪv/ a. 有助（于…）的
commercialize /kəˈmɜːʃəˌlaɪz/ v. 使商业化
density /ˈdensəti/ n. 密度
infancy /ˈɪnfənsi/ n. 初期
onshore /ˈɒnˌʃɔː/ a. 在陆地上的；在岸上的
executive /ɪɡˈzekjʊtɪv/ a. 执行的；行政的
commonwealth /ˈkɒmənwelθ/ n. 联邦；联合体
opposition /ˌɒpəˈzɪʃn/ n. 反抗；对抗
reasonably /ˈriːznəbli/ ad. 相当地
furthermore /ˈfɜːðəmɔː/ ad. 此外；而且
specification /ˌspesɪfɪˈkeɪʃn/ n. 规格
rig /rɪɡ/ n. 钻井架；钻塔
entitle /ɪnˈtaɪtl/ v. 给予…权利（或资格）

royalty /ˈrɔɪəlti/ n. 使用费
temper /ˈtempə/ v. 使缓和；使温和
landholder /ˈlændˌhəʊldə/ n. 土地所有者
compensation /ˌkɒmpənˈseɪʃn/ n. 赔偿金；补偿金
corporate /ˈkɔːp(ə)rət/ a. 公司的；团体的
advisory /ədˈvaɪz(ə)ri/ a. 顾问的；咨询的
acreage /ˈeɪkərɪdʒ/ n. 以英亩计算的面积
proposition /ˌprɒpəˈzɪʃn/ n. 提议；建议
necessarily /ˈnesəsərəli; ˌnesəˈserəli/ ad. 必定；必然
clarity /ˈklærəti/ n. 清晰；明晰
indication /ˌɪndɪˈkeɪʃn/ n. 迹象；标示

▶ Text understanding

1 Read Paras. 1-5 and complete the summary of these paragraphs.

Australia is home to the 1) _____ largest recoverable shale gas reserves in the world. It has several favorable factors for commercializing the resource. However, the Australian shale gas industry is still in its 2) _____. In the past few years, exploration activities in Australia have increased. Big companies like Chevron, ConocoPhillips, Statoil, Total, BG Group, have invested over 3) _____ in Australia's shale gas industry as of mid-2013, according to the EIA.

2 What are the challenges that the commercialization of the shale gas resource in Australia faces. Read Paras. 6-13 and complete the following table.

Challenges	Supporting details
Higher production cost	Several basins are located in 1) _____ parts of the country, which would require significant investments in pipeline and processing 2) _____.
Underdeveloped service industry	The capacity of Australia's service industry in terms of 3) _____ and hydraulic fracturing services is not 4) _____. As a result, shale gas drilling and fracking are over 5) _____ times as costly as in the US.
Unfavorable regulatory environment	Key challenge in the regulatory environment is the mineral rights 6) _____. In Australia, the state owns any underground resources. So there is less 7) _____ for landholders to encourage minerals development on their properties.
Uncertain quality of shale resource	Reserves of shale gas haven't been 8) _____ in a big way, making investing in the sector a high risk 9) _____.

3 What suggestions are given to those interested in investing in Australian shale resource?

_____

_____

▶ Language building

1 Match each of the words in the left column with its corresponding meaning in the right column.

_____	1 recoverable	a to develop sth. so that you can sell it and make a profit
_____	2 conducive	b creating a situation that helps sth. to happen
_____	3 commercialize	c capable of being regained
_____	4 know-how	d a comparison between two things that have similar features
_____	5 density	e the degree to which an area is filled with people or things
_____	6 analogy	f knowledge, practical ability or skill to do sth.
_____	7 infrastructure	g far from towns or other places where people live
_____	8 infancy	h the money that you receive when you sell sth.
_____	9 proceeds	i the time when sth. is just starting to be developed
_____	10 remote	j basic systems and structures that a country or organization needs in order to work properly

128 | Unit 5

2 Among the three choices given, choose the one that is **NOT** close in meaning to the underlined word in each sentence.

1. This has ensured local communities are able to benefit financially, thus helping to temper local opposition to the industry.
   A) weaken   B) reproduce   C) lessen
2. Shale gas drilling and fracking are over three times as costly as in the US.
   A) pricey   B) expensive   C) cheap
3. The proceeds are generally not as large as US royalty payments.
   A) advantages   B) profits   C) gains
4. Finally, there remains a great deal of uncertainty around the quality of Australia's shale resource as reserves haven't been tapped in a big way yet.
   A) stocks   B) reservoirs   C) reverses
5. There is less incentive for landholders to encourage minerals development on their properties.
   A) emphasis   B) encouragement   C) motivation
6. In the US, landowners possess the rights to the resources beneath their land and are entitled to royalties.
   A) qualified   B) eligible   C) crowned
7. This year, some clarity will begin to emerge as a number of companies will publish flow test results from their shale wells.
   A) appear   B) merge   C) surface
8. The US has a unique system where landowners own the minerals, so there's a win-win.
   A) unusual   B) distinct   C) fancy

3 Complete each of the following sentences with an appropriate word from the given word family. Change the form where necessary.

1. Most analysts believe the country's _____ reserves are well over 100 billion barrels. (recoverable, recovery, recovered)
2. Even in the _____ heart of its largest city, religion remains central to life in Myanmar. (commerce, commercial, commercialize)
3. Heavy fogs in the downtown area are the mixture of natural fog, smoke and dust, and the river valley and urban heat island effects _____ fogs. (dense, density, densify)
4. At the moment, the _____ on how much human DNA can be put into an animal is vague. (regulate, regulation, regulatory)
5. Some diseases are _____ from one generation to the next. (transmit, transmission, transmittance)
6. They both have the _____ of winning. (capable, capacity, capability)
7. The reasons include political instability and an _____ level of education in several countries. (adequate, adequacy, inadequate)

8 If you are not prepared, then you will not be able to _____ in the class discussion. (participate, participation, participant)

9 They recently ran a series of tests to measure the _____ of the drug. (effective, effect, efficacy)

10 Are you for or against this _____? (propose, proposal, proposer)

4 Match each word in the box with the group of phrases where it is usually found.

| drop | risk | launch | deal | tap | drill |

1 _____
~ off
a ~ in price
a dramatic ~
a sharp ~

2 _____
a great ~ of
a good ~ of
a ~ of
a big ~

3 _____
~ deep into
~ a hole
~ for oil
~ through

4 _____
~ a product
~ new drugs
~ a service
~ a novel

5 _____
high ~
~ management
at ~
face the ~ of

6 _____
~ reserves
~ into
~ the expertise of
~ energy

5 Find the idiomatic expressions in the text matching the Chinese equivalents.

1 处在起步阶段 _____
2 生产成本 _____
3 矿产产权归属 _____
4 监管环境 _____
5 引起国际关注 _____
6 行业参与者 _____
7 国家的偏远地区 _____
8 高风险提议 _____
9 为时过早 _____
10 可采储备 _____
11 页岩富集地区 _____
12 现有基础设施 _____
13 低人口密度 _____
14 全球第四大 _____
15 全球参与者 _____
16 最相似的例子 _____
17 有潜力 _____
18 首先 _____
19 高规格 _____
20 对…享有权利 _____

6 Combine the sentences given below. Then compare your sentences with the original ones in the text.

1 _____

    a   Australia is home to the world's seventh largest recoverable shale gas reserves.
    b   Australia has several characteristics conducive for commercializing shale gas.
    c   These characteristics include existing infrastructure, industry know-how and low population density in shale-rich regions.

2 _____

    a   Australia has an estimated 437 Tcf of recoverable shale gas reserves, according to the EIA.
    b   The US has an estimated 665 Tcf of recoverable shale gas reserves and ranks the world's fourth largest.
    c   Australian recoverable shale gas reserves are around two-thirds of the US.

3 _____

    a   Australia's shale sector holds promise.
    b   Its challenges are very different to the US.
    c   Production costs, the onshore service industry and the regulatory environment are different between Australia and the US.
    d   These differences could constrain the industry's potential.

4 _____

    a   This has ensured local communities are able to benefit financially.
    b   So it helps to temper local opposition to the industry.

5 _____

    a   The proceeds are generally not as large as US royalty payments.
    b   This information is from RFC Ambrian.
    c   RFC Ambrian is a firm providing corporate financial advisory services in the natural resources market.

7 Translate each of the Chinese sentences into English by using the underlined phrase or structure in the example.

1. Australia, <u>home to</u> the world's seventh largest recoverable shale gas reserves, has several characteristics conducive for commercializing the resource.
中国有 14 亿人口。
_____

2. Broadly most people would recognize that Australia is the closest <u>analogy to</u> the US, with our infrastructure position and unconventional shale gas opportunities.
计算机的工作程序和人脑的运作有着惊人的相似度。
_____

3. The Australian shale gas industry is still <u>in its infancy</u>, but exploration has increased in the last few years.
该岛的旅游业发展在很大程度上仍处于初期阶段。
_____

4. While Australia's shale sector <u>holds promise</u>, its challenges are very different to the US in terms of production costs, the onshore service industry and the regulatory environment.
本课程还将向学生介绍一些虽未经测试，但相信在不久的将来会投入使用的材料。
_____

5. <u>For starters</u>, developing shale resources in Australia is costlier than in the US.
首先，这些国家不太可能拿本国大规模的外汇储备冒险。
_____

6. Landowners possess <u>the rights to</u> the resources beneath their land and are entitled to royalties
去年年底，我买下了这部小说的版权。
_____

7. This has ensured local communities are able to benefit financially, <u>thus helping to</u> temper local opposition to the industry.
这些措施将降低粮食生产过程中的浪费，从而有助于保持粮食价格稳定。
_____

Academic writing

▶ Micro-skill: Generalizations

To introduce a topic, generalizations are often used. They are often simple and easy to understand. However, generalizations must not be inaccurate or too simplistic.

1 The use of generalizations

Compare the following two sentences.

1 There is an increasing number of Chinese students going abroad to study.
2 The number of students going abroad to study increased from 23,000 to 123,000. About 56.2% of them went to the United Sates.

The first sentence is easier to understand and remember, while the second one provides more accurate data. Generalizations can be used for a simple picture of a topic when accuracy is not necessary.

Generalizations can be valid and invalid. While valid generalizations can always be justified, invalid generalizations are the ones that cannot be supported by evidence or research. For example:

Men are smarter than women. (Invalid generalization)
Internet has changed the way we live. (Valid generalization)

1 Decide which of the followings are valid generalizations.

1 People in North China are taller than those in South China.

2 A tsunami is difficult to predict.

3 There is a link between poverty and crime.

4 Girls study harder than boys.

5 Air travel is faster than train travel.

2   Ways for expressing generalizations

There are two ways to express generalizations.

1) The plural (more common):

   Books are the best friends of human beings.

2) The definite article + the singular (more formal):

   The book is the best friend of human beings.

Note: Avoid generalizations that are absolute, such as:

   Children learn things much faster than adults do.

Absolute generalizations are often invalid because there may well be exceptions. A more cautious generalization could be:

   Children tend to learn things much faster than adults do.

2   Read the following passage and underline the generalizations.

The loss of wildlife and plants in the UK shows no sign of slowing. This year's *State of Nature*, the most comprehensive assessment yet, found that the area occupied by more than 6,500 species has shrunk by 5% since 1970. Of the species with detailed data, nearly 700 saw numbers fall by 13%. While the 5% fall in distribution of species may sound small, Hayhow calls it a "canary in the coalmine signal" because changes in distribution usually happen much more slowly than changes in abundance of wildlife. This year's report was, for the first time, able to draw on figures on less well-studied species, such as lacewings, hoverflies and lichens, after biases in the data were adjusted for. For moths and butterflies, the picture is one of steep declines. Mammals and birds show a slight increase since 1970, which masks dramatic falls in some species such as hen harriers. There is a flicker of good news for hedgehogs – classed as vulnerable to extinction due to long-term decline, since 2012 their numbers seem to have grown. The biggest drivers for change are intensified farming and climate change.

3   Write generalizations on the following topics.

1   fertile land / good crop yields

_____

2   honest performance in class / respect for the teacher

_____

3   hard working / academic success

_____

4   smoking / health

5   good relationship / mutual understanding

4 Study the table and write five generalizations using the information.

The table shows results of a college survey on what students prefer to do during the summer vacation.

	Undergraduates (%)		Graduates (%)	
	Male	Female	Male	Female
Traveling	21	18	48	33
Doing part-time jobs	20	29	27	39
Looking for an internship	14	19	13	15
Doing voluntary work	14	23	5	9
Staying on campus	6	9	4	2
Other	13	10	5	4

1 _____
2 _____
3 _____
4 _____
5 _____

Macro-skill: Classification

To classify is to sort things into several classes, types or categories according to their characteristics or features. One group of items may be classified in different ways based on different criteria. For instance, people can be classified into men and women by gender; young, middle-aged and old by age; wealthy, middle class and poor by income, etc. To find out the makeup of the students of a school, the school office may classify the students according to their grades and other aspects of academic performance if the purpose is to evaluate the results of teaching. The office may also group the students according to annual family income or sources of income if the purpose is to determine how many of them need financial aid. Or they might classify them on the basis of their interest in extracurricular activities if they are considering the establishment of student clubs.

Whatever the purpose and criteria may be, the categories used must be mutually exclusive and must involve all the items to be classified. For example, while it would be confusing to classify students as boys, girls and athletes, classifying them just according to sports teams would not account for those who have not joined any team.

1 Read the following passage and complete the table.

There are many different types of crude oils and natural gases, some more valuable than others. Heavy crude oils are very thick and viscous and are difficult or impossible to produce. Light crude oils are very fluid, relatively easy to produce, rich in gasoline, and more valuable. Some natural gases burn with more heat than others and are more valuable. Sulfur is a bad impurity in both natural gas and crude oil. Crude oils are classified as sweet and sour on the basis of their sulfur content. Sweet crudes have less than 1% sulfur by weight, whereas sour crudes have more than 1% sulfur.

Objects to be classified (Topic)			1) _____
Classification	Crude oils	By thickness	2) _____ and 3) _____
		By sulfur content	4) _____ and 5) _____
	Natural gases	By release of heat	6) _____ and 7) _____

2 Read the following passage and complete the table.

It is well known that science plays an important role in the societies in which we live. Many people believe, however, that our progress depends on two aspects of science. The first is the application of the machines, products and systems of applied knowledge that scientists and technologists develop. Through technology, science improves the structure of society and helps man to gain increasing control over his environment. The second aspect is the application by all members of society from the government official to the ordinary citizen, of the special methods of thought and action that scientists use in their work.

Objects to be classified (Topic)	1)
Classification	2)
	3)

3 Read the following passage and complete the table.

People are bound within relationships by two types of bonds: expressive ties and instrumental ties. Expressive ties are social links formed when we emotionally invest ourselves in and commit ourselves to other people. Through association with people who are meaningful to us, we achieve a sense of security, love, acceptance, companionship and personal worth. Instrumental ties are social links formed when we cooperate with other people to achieve some goal. Occasionally, this may mean working with instead of against competitors. More often, we simply cooperate with others to reach some end without endowing the relationship with any larger significance.

Sociologists have built on the distinction between expressive and instrumental ties to distinguish between two types of groups: primary and secondary. A primary group involves two or more people who enjoy a direct, intimate, cohesive relationship with one another. Expressive ties predominate in primary groups; we view the people as ends in themselves and valuable in their own right. A secondary group entails two or more people who are involved in an impersonal relationship and have come together for a specific, practical purpose. Instrumental ties predominate in secondary groups; we perceive people as means to ends rather than as ends in their own right. Sometimes primary group relationships evolve out of secondary group relationships. This happens in many work settings. People on the job often develop close relationships with coworkers as they come to share gripes, jokes, gossip and satisfactions.

Objects to be classified (Topic)	1) _____
Classification	2) _____ predominate in 3) _____ groups. 4) _____ predominate in 5) _____ groups. In workplaces, 6) _____ may evolve into 7) _____.

Writing assignment

Choose one of the topics below and compose a short essay of about 120 words following the organization of classification.

1 Major types of modern energy resources

2 Practical problems of wind and solar power

Sharing

You attend the Energy Science Conference of China on behalf of your university and are to make an analysis report on China's potential to reach peak carbon emissions and carbon neutrality in terms of developing renewables and shale gas.

1 Work in groups to search for information about the energy transformation and development of renewables and shale gas in China, including relevant statistics.

2 Summarize what you have obtained from your investigation, and discuss with your group members the potential of China to develop renewables and shale gas and your prediction of China's future energy transformation related to peak carbon emissions and carbon neutrality.

3 Deliver an oral report with PPT in class. Your report is expected to include the following parts and employ statistics.

1) A brief introduction to China's goals of attaining peak carbon emissions and carbon neutrality and energy transformation in China
2) Analysis of China's potential to develop renewables
3) Analysis of China's potential to develop shale gas
4) Suggestions on China's energy transformation to achieve the goals

Unit 6 Energy and Climate Change

Lead-in

Climate change is mainly caused by energy consumption. Human beings are to blame for their various activities that release greenhouse gas emissions which eventually lead to climate consequences. Thus it is also human beings' responsibility to reduce climate change. Then, what can we do?

To tackle climate change, the Paris Agreement sets a long-term goal to hold the increase in the global average temperature to well below 2℃ above pre-industrial levels. However, countries find it hard to enforce the agreement since there are no detailed procedures or criteria. Some steps are then proposed to make this process more feasible. Do you know what they are?

Learning objectives

Upon completion of this unit, you will be able to:
- talk about the specific measures that can be taken to mitigate climate change;
- discuss the basics of the 2℃ goal of COP21 agreement;
- use argument and discussion in academic writing;
- write a conclusion or summary for an essay;
- compose an energy development plan for your city in response to climate change.

Viewing and speaking

▶ Viewing

New words

option /ˈɒpʃn/ n. 选择；选项
capture /ˈkæptʃə/ n. & v. 捕获

aviation /ˌeɪviˈeɪʃn/ n. 航空工业
volatility /ˌvɒləˈtɪləti/ n. 易变性

1 Watch the video clip and choose the best answer to each of the following questions.

1. How much will the UK reduce greenhouse gas emissions by 2050?
   A) By 20%.   B) By 50%.   C) By 80%.   D) By 100%.
2. Which is NOT one of the three largest causes of emissions in the UK?
   A) Agriculture.   B) Transport.   C) Heating.   D) Electricity.
3. Which is NOT one of the three technologies to provide a low-carbon solution?
   A) Converting sustainable natural resources into usable energy.
   B) Importing energy from overseas.
   C) Carbon capture and storage technology.
   D) Using nuclear physics to generate electricity.
4. What is low-carbon energy featured by?
   A) It's rare.   B) It's plentiful.   C) It's secure.   D) It's dangerous.

2 Watch the video clip again and fill in the blanks with what you hear.

1. Human _____ has made it possible to harness energy and use it to power our society, providing heat and electricity to buildings, _____ us from A to B, and fueling our industry.
2. Burning fossil fuels to harness energy causes greenhouse gas _____ that drive global warming. Scientists agree that a _____ rise of more than two degrees would have an unexpected impact on our lives.
3. Electricity is just one part of a _____ picture. Across aviation, industry, transport, heating, agriculture and _____, there are also choices to make.
4. As a nation, we need to decide how we're going to _____ ourselves and future generations to come. The decisions we make are not just _____ ones.
5. They will _____ on all areas of life: from the economy and employment to health and security. The carbon plan is the government's _____ to this challenge.

Energy and Climate Change | 141

Speaking

1 Read the short passage aloud and pay attention to the idea about climate change.

The global climate is changing and posing increasingly severe risks for ecosystems, human health and the economy. The earth is already facing impacts of a changing climate, including rising sea levels, more extreme weather, flooding, droughts and storms.

The changes are happening because large amounts of greenhouse gases are released into the atmosphere as a result of human activities worldwide. About two-thirds of global greenhouse gas emissions are linked to burning fossil fuels for energy to be used for heating, electricity, transport and industry. Combustion of fossil fuels also releases air pollutants that harm the environment and human health.

Mitigating and adapting to climate change are key challenges of the 21st century. To succeed in limiting global warming, the world urgently needs to use energy efficiently while embracing clean energy sources.

2 Work in pairs and answer the two questions.

1 What is the reason for climate change according to this passage?
_____
_____

2 What can we do to mitigate climate change?
_____
_____

3 Do some research on climate change, a term you will find in the texts in this unit, and then prepare a one-minute oral presentation. Make sure the messages are properly delivered to the audience.

Extensive reading

Reading text

What We Can Do to Reduce Climate Change

1. What are the causes of climate change? We are. While a wide range of natural phenomena can radically affect the climate, climate scientists overwhelmingly agree that global warming and resultant climate effects that we're witnessing are the result of human activities. Because climate change today is very much linked to global warming, efforts to fight climate change are very much about reducing the greenhouse gases in the atmosphere.

Cleaner energy sources

2. One important way to fight climate change is to use alternative sources of renewable energy which will not release harmful emissions. Some of these cleaner sources of renewable energy include wind energy, solar energy, water or hydropower, biomass and geothermal energy.

3. By reducing our reliance on and usage of fossil fuels, and tapping into alternative and greener sources of energy, not only are we helping reduce the release of greenhouse gases (especially carbon dioxide) into the atmosphere, and hence helping reduce global warming and fight climate change, we are also helping ensure the sustainability of the world's development.

Reforestation

4. The cleanest and most efficient remover of carbon dioxide from our atmosphere actually comes free. This remover is a gift from nature – our green plants and trees.

5. Unfortunately, we have taken this gift for granted. The rate at which we are cutting down our trees and forests to make way for human development has greatly reduced the earth's ability to remove carbon dioxide from the atmosphere. This has in turn contributed to a faster rate of global warming and climate change.

6. You can do your part by contributing to reforestation efforts. Take part in your local community's plant-a-tree effort. Grow an organic garden of your own. Reduce your usage of paper – paper production is one main cause of deforestation. But if you must, use recycled paper.

Organic farming

7. Soils are an important sink for atmospheric carbon dioxide. Nevertheless, deforestation to make way for conventional agriculture is increasingly depleting this sink.

8. Sustainable and organic agriculture helps counteract climate change by restoring soil organic matter content as well as reducing soil erosion and improving soil physical structure. Organic farming also does not use chemical fertilizers that release substantial nitrous oxide and methane (greenhouse gases) into the environment, and as such reduces global warming, while at the same time maintaining crop yields.

9. As individuals, to reduce climate change through organic farming is to switch to organic products. As the demand for organically grown products increases, organic farming will become more economically viable and more popular. Or better still, grow your own organic garden. As a community, encouraging local farmers to cultivate various products by rotating their crops can also go a long way in reducing carbon emissions and restoring the ecological balance in the environment.

Green driving

10. Given the number of automobiles in the world today, they emit a substantial amount of greenhouse gases and contribute significantly to global warming and climate change. In fact, after coal-burning power plants, automobiles are the second largest source of carbon dioxide, which is why cutting greenhouse gas emissions from automobiles is a critical strategy to fight global warming and climate change.

11. The best strategy on how to reduce climate change is definitely to reduce the use of automobiles. Use public transport or carpool if you can, instead of driving your own car. This is because the amount of pollution created when you drive by yourself (as opposed to carpooling) can be as bad as taking the plane! If you must drive, then use an eco-friendly and fuel efficient car, and ensure that your vehicle's engine is working in impeccable condition at all times. A car in poor working condition uses more fuel for the same mileage covered, and releases more carbon dioxide and exhaust gases.

Green shopping

12 One point on how to reduce climate change through green shopping is to buy local products instead of those produced overseas.

13 Transporting exotic fruits and vegetables from one destination to another requires a lot of energy, usually from the burning of fossil fuels, which contributes to greenhouse gas emissions and global warming. Moreover, in attempts to keep the fruits and vegetables fresh, chemical pesticides and preservatives are used, which again contribute to toxins and greenhouse gases in the atmosphere.

14 Hence, buy local products if you can – not only do they leave less carbon footprints and are cheaper, they are also healthier. And reduce consumption, reuse whenever you can, and recycle whatever you no longer need. Then complete the recycling loop by buying recycled products.

Reduce-Reuse-Recycle practices

15 The culture of consumerism today encourages people to buy and throw, with little consideration for the impact of such unsustainable consumption on the world. For each item that we purchase and use, energy and resources are used in its manufacture, packaging, transportation and retail, and ultimately its disposal. Pollution is created at each step of the process, and substantial greenhouse gases are also released. It is time we should think twice about the way we live.

16 If we are to live a sustainable and green life, leaving minimal carbon footpaths during our passage through this earth, we need to resist the culture of consumerism, and instead adopt a lifestyle guided by the Reduce-Reuse-Recycle principle. Reducing, reusing and recycling help us conserve resources and energy, and reduce pollution and greenhouse gas emissions produced, for example in the raw material extraction and disposal processes. In turn, reducing our greenhouse gas emissions would substantially help us reduce global warming and fight climate change.

Start now

17 It is time that we should make efforts for a green life, reduce our carbon footprints and fight climate change. We need to start making concentrated efforts at reducing consumption and waste and put recycling into practice. And we can help undo some of the impact of deforestation by contributing to forest conservation and growing efforts. We need to conserve energy and resources, through using less electricity, as well as investing in energy efficient appliances. We can invest in greener sources of energy, such as solar energy. We can also adopt green driving tips and green living practices. We also need to switch to more organic methods of agriculture and seek more sustainable methods. We can also help reduce pollution by using more eco-friendly products.

Culture notes

organic farming: It is the agricultural system that uses ecologically based pest controls and biological fertilizers derived largely from animal and plant wastes and nitrogen-fixing cover crops. Modern organic farming was developed as a response to the environmental harm caused by the use of chemical pesticides and synthetic fertilizers in conventional agriculture, and it has numerous ecological benefits.

rotate crops: Crop rotation is the successive cultivation of different crops in a specified order on the same fields, in contrast to a one-crop system or to haphazard crop successions. It is done so that the soil of farms is not used for only one set of nutrients. It helps in reducing soil erosion and increasing soil fertility and crop yields.

Vocabulary

overwhelmingly /ˌəʊvəˈwelmɪŋli/ ad. 压倒性地
resultant /rɪˈzʌltənt/ a. 由此引起的
witness /ˈwɪtnəs/ v. 目击；目睹
reforestation /ˌriːfɒrɪˈsteɪʃn/ n. 重新造林
remover /rɪˈmuːvə/ n. 清除剂
sink /sɪŋk/ n. 水池
atmospheric /ˌætməsˈferɪk/ a. 大气层的
counteract /ˌkaʊntərˈækt/ v. 抵消
erosion /ɪˈrəʊʒn/ n. 侵蚀；腐蚀
yield /jiːld/ n. 产量
switch /swɪtʃ/ v. 使转变；使转换
viable /ˈvaɪəbl/ a. 切实可行的
rotate /rəʊˈteɪt/ v. 使轮换
impeccable /ɪmˈpekəbl/ a. 完美的；无瑕疵的

mileage /ˈmaɪlɪdʒ/ n. 英里里程
exhaust /ɪɡˈzɔːst/ n. 尾气；废气
exotic /ɪɡˈzɒtɪk/ a. 异国的；外来的
preservative /prɪˈzɜːvətɪv/ n. 防腐剂；保鲜剂
toxin /ˈtɒksɪn/ n. 毒素
loop /luːp/ n. 环；圈
consumerism /kənˈsjuːməˌrɪz(ə)m/ n. 消费主义
ultimately /ˈʌltɪmətli/ ad. 最后；最终
disposal /dɪˈspəʊzl/ n. 清除；处理
minimal /ˈmɪnɪml/ a. 极小的；极少的
passage /ˈpæsɪdʒ/ n. 通过；经过

Integrated exercises

1 Match each of the following statements with the paragraph from which the information is derived. You may choose a paragraph more than once.

_____ 1 Buying organic products will encourage the farming of organic food.

_____ 2 Coal-fired power plants are the number one emitter of carbon dioxide, followed by automobiles.

_____ 3 Green plants and trees can help clean up the carbon dioxide in the atmosphere.

_____ 4 Organic farming has many benefits for air quality, soil quality and crop yields.

_____ 5 People care too little about plants and trees, and deforestation has sped up global warming and climate change.

_____ 6 People today tend to be extravagant in buying and throwing products under the influence of consumerism.

_____ 7 Shipping fresh foods from overseas consumes large amounts of fossil fuels and more chemical pesticides.

_____ 8 The working condition of a car also affects the amount of fuel consumed per mile.

_____ 9 There is a wide consensus among climate scientists that it is human activities that lead to global warming and climate change.

_____ 10 Using clean energy can contribute to the sustainable development of the earth.

2 Complete the following table to check your understanding of the major points and structure of the text.

Introduction (Para. 1)	Climate scientists 1) _____ agree that global warming and climate effects are the result of 2) _____, and efforts should be made to reduce the greenhouse gases in the atmosphere.
Body (Paras. 2-16)	• **Cleaner energy sources** Using cleaner energy will reduce the release of 3) _____ and ensure the 4) _____ of the world. • **Reforestation** Deforestation has greatly reduced the earth's ability to 5) _____ from the atmosphere. What we can do is to contribute to reforestation, grow an organic garden, and reduce the 6) _____. • **Organic farming** Organic agriculture restores soil organic matter content, reduces 7) _____ and improves soil physical structure. Individuals should switch to 8) _____ or to grow their own organic garden.

(to be continued)

Body (Paras. 2-16)	(continued) • **Green driving** Automobiles are the second largest source of carbon dioxide, so it is better to use public transport or 9) _____ than driving by yourself. And choose an eco-friendly and 10) _____ whose engine is in impeccable condition. • **Green shopping** Buying local products saves the energy consumed during the transportation of 11) _____ fruits and vegetables, as well as the 12) _____ and preservatives to keep them fresh. • **Reduce-Reuse-Recycle practices** For each item, pollution is created in its manufacture, packaging, transportation, retail and 13) _____. We need to resist the culture of 14) _____, and adopt a lifestyle guided by the Reduce-Reuse-Recycle principle.
Conclusion (Para. 17)	It is time for us to make efforts for a 15) _____, reduce our carbon footprints and fight climate change.

3 Work in pairs and list the measures that can be taken to mitigate climate change. Take notes of the measures while you do this task.

_____
_____
_____

4 Translate the phrases into Chinese.

1 make way for _____
2 do one's part _____
3 organic farming _____
4 conventional agriculture _____
5 soil erosion _____
6 chemical fertilizer _____
7 crop yield _____
8 switch to _____
9 go a long way _____
10 restore the ecological balance _____

11 as opposed to _____
12 fuel efficient car _____
13 in impeccable condition _____
14 exhaust gases _____
15 the culture of consumerism _____
16 at all times _____
17 think twice _____
18 put ... into practice _____
19 undo the impact _____
20 forest conservation _____

5 Complete the sentences by translating the Chinese given in brackets into English. Refer to the phrases listed above when necessary.

1 His proposition is _____ (在理论上可以，但不能付诸实践).
2 How does a poem change when you _____ (大声读出来而不是将它写在纸上)?
3 Several houses were _____ (为了建新路而被拆除).
4 I advised him to _____ (三思之后再决定是否退学).
5 That sort of approach should _____ (对打破僵局大有帮助).

Intensive reading

▶ Warming up

1 We talk a lot about climate change. What is climate change and what is the cause of it? The following passage may help clarify these questions.

Climate change is a change in the statistical distribution of weather patterns when that change lasts for an extended period of time (i.e., decades to millions of years). Climate change may refer to a change in average weather conditions, or in the time variation of weather within the context of longer-term average conditions. Climate change is caused by factors such as biotic processes, variations in solar radiation received by the earth, plate tectonics, and volcanic eruptions. Certain human activities have been identified as primary causes of ongoing climate change, often referred to as global warming.

Scientists actively work to understand past and future climate by using observations and theoretical models. A climate record – extending deep into the earth's past – has been assembled, and continues to be built up, based on geological evidence from borehole temperature profiles, cores removed from deep accumulations of ice, floral and faunal records, glacial and periglacial processes, stable-isotope and other analyses of sediment layers, and records of past sea levels. More recent data are provided by the instrumental record. General circulation models, based on the physical sciences, are often used in theoretical approaches to match past climate data, make future projections, and link causes and effects in climate change.

2 Work in groups and discuss the consequences of climate change and how to deal with them.

▶ Reading text

1 The international community has adopted a common goal for action on climate change – the 2°C target – in the UNFCCC. While this is a valid long-term goal, it is not easy to translate it into practical policy and investment needs. Further steps are needed in the COP21 agreement to lock in the 2°C vision more strongly, so that it anchors future expectations, guiding policymaking and both public and private energy sector investment decisions, and acts as a standard against which short-term government targets and actions can be assessed. These further steps form the third key pillar of our recommended approach to the COP21 agreement.

2 Failure to link short- and long-term decisions is costly. Many energy sector investment decisions relate to long-lived capital-intensive infrastructure, emphasizing the need to make effective risk assessments about future developments. A focus only on short-term emissions targets to 2025, themselves important to avoid excess emissions in the short term, could lead to the adoption of a technology mix that is far from optimal if not devised in the light of a longer-term decarbonization plan, resulting in increasing the cost of achieving critical climate goals.

Locking in the Long-term Vision

3. The first step to lock in the 2°C vision more strongly in the COP21 agreement would be to re-express that goal in terms of a long-term greenhouse gas emissions target. This would be more straightforward to apply in the energy sector and lend itself to easier accounting for monitoring. A number of possible options for supplementing the 2°C goal in this way have been tabled for consideration in the run-up to COP21, including a target level for 2050 emissions, a target date for net-zero emissions in the second half of this century, or a timeline for the phaseout of unabated fossil fuels. Adoption of such specific, emissions-focused long-term goals to complement the existing 2°C goal would be a valuable additional signal to the energy sector of the need for transformative change.*

4. The second step, needed to ensure that any long-term goal is meaningful – whether framed in terms of temperature or emissions – is to create a clear link between it and countries' actions, rather than it being just an aspirational statement. The global goal should be used by all countries to inform nationally determined decarbonization or low-carbon development pathways, which would in turn provide an important benchmark for the short-term goals countries set in the five-year review cycle. Whether the national pathway involves an immediate fall in total emissions (such as in the United Kingdom's carbon budget framework) or a peak then later decline (such as in South Africa's pathway) will usually depend on the country's development status. Many developing countries give priority to their economic and social development. The COP21 agreement must recognize this, embedding climate goals in plans which permit countries to achieve these aspirations.

5. As well as assisting in sending a clear signal to investors in new long-lived assets, the newly expressed long-term goal and low-carbon development pathways will shape decisions on the operation or retirement of existing capital stock and technology development. Goals and actions reflected in the INDC Scenario result in operating existing power plants and those under construction in such a way that they emit some 275 Gt from 2015 to 2040 (Figure 1). The efficiency measures and the push toward renewables adopted in the Bridge Scenario would reduce this by almost 10%; but a trajectory compatible with a 2°C objective that fully internalizes a long-term decarbonization vision would need to cut

cumulative emissions by around 35% by driving action to ensure the retirement of old high-emitting capital stock, allowing only very efficient or low-carbon generating plants to operate.

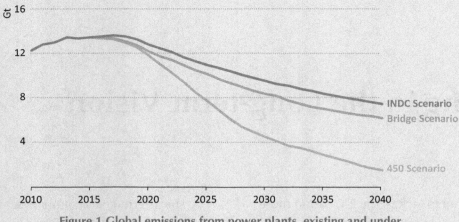

Figure 1 Global emissions from power plants, existing and under construction

6 A long-term global emissions goal can also boost further energy technology development. The benefits of earlier investment in solar and wind generation are now being seen in rapidly falling prices. The same process will need to be followed to bring forward carbon capture and storage (CCS), battery storage, electric and other low-emissions vehicles and other emerging technologies (Figure 2). A clear, measurable long-term vision in the COP21 agreement will underpin investment in those key technologies that are essential to unlocking long-term emission reductions compatible with a 2°C trajectory.

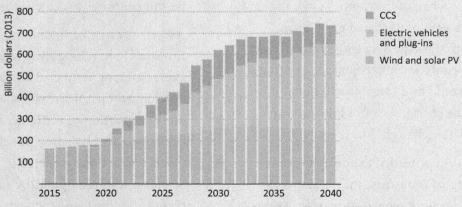

Figure 2 Global investment in variable renewables, CCS and electric vehicles in the 450 Scenario

7 There are also a number of more detailed ways the COP21 decisions (and subsequent technical work programs) can address aspects of technology development: reporting on technology progress ahead of each five-year review; global goals for research, development, demonstration and deployment (RDD & D) levels; country reporting on low-carbon technology actions; and strengthening the link between the technology and finance aspects of the UNFCCC. The technology needs assessments and

technology action plans prepared by developing countries should be integrated into wider low-carbon development strategies, improving alignment between these countries' development, mitigation and technology goals. Parties should also be encouraged to increase international cooperation to scale up low-carbon technologies. There are many existing multilateral technology and policy collaboration on topics such as carbon pricing, renewable energy and energy efficiency. Care should be taken to build on, consolidate and not duplicate these efforts.

* The particular form of the goal would need to take into account the uncertainty in translating temperature rise probabilities into specific emissions targets and the need for flexibility to incorporate updated scientific knowledge.

Culture notes

UNFCCC (The United Nations Framework Convention on Climate Change): It is an international environmental treaty adopted in 1992. Its stated objective is to achieve stabilization of greenhouse gas concentrations in the atmosphere at a low enough level to prevent dangerous anthropogenic interference with the climate system.

COP21: It is also called the Paris Agreement and was adopted in December 2015, which aimed to reduce the emission of gases that contribute to global warming. The Paris Agreement set out to improve upon and replace the Kyoto Protocol, an earlier international treaty designed to curb the release of greenhouse gases. The objective was no less than a binding and universal agreement designed to limit greenhouse gas emissions to levels that would prevent global temperatures from increasing more than 2°C above the temperature benchmark set before the beginning of the Industrial Revolution.

INDC (Intended Nationally Determined Contributions): In advance of the COP21, countries publicly put forward their intended post-2020 climate action plans, which help lay a foundation for their actions to address climate change, including the mitigation of greenhouse gas emissions.

Vocabulary

lock in 固定；锁定
vision /'vɪʒn/ n. 构想；设想
policymaking /'pɒləsiˌmeɪkɪŋ/ n. 决策
pillar /'pɪlə/ n. 支柱
long-lived /ˌlɒŋ'lɪvd/ a. 长久的
adoption /ə'dɒpʃn/ n. 采用；采纳
optimal /'ɒptɪml/ a. 最佳的
decarbonization /diːˌkɑːbənaɪ'zeɪʃn/ n. 脱碳
supplement /'sʌplɪˌment/ v. 增补；补充
run-up /'rʌnʌp/ n. 预备阶段
timeline /'taɪmlaɪn/ n. 时间轴；年表
phaseout /'feɪzaʊt/ n. 逐步淘汰
unabated /ˌʌnə'beɪtɪd/ a. 不减弱的；不衰退的
complement /'kɒmplɪˌment/ v. 补充
transformative /træns'fɔːmətɪv/ a. 彻底改观的
aspirational /ˌæspɪ'reɪʃ(ə)nl/ a. 有雄心壮志的

pathway /'pɑːθˌweɪ/ n. 途径；路径
benchmark /'bentʃˌmɑːk/ n. 基准
embed /ɪm'bed/ v. 使融入
asset /'æset/ n. 资产
scenario /sə'nɑːriəʊ/ n. 方案；情节；剧本
compatible /kəm'pætəbl/ a. 相容的
internalize /ɪn'tɜːnlaɪz/ v. 使内化
cumulative /'kjuːmjʊlətɪv/ a. 累积的
underpin /ˌʌndə'pɪn/ v. 巩固；支持
unlock /ʌn'lɒk/ v. 提供；开启
alignment /ə'laɪnmənt/ n. 支持；结盟
scale up 扩大；增加
multilateral /ˌmʌlti'læt(ə)rəl/ a. 多边的；多国的
collaboration /kəˌlæbə'reɪʃn/ n. 合作；协作
consolidate /kən'sɒlɪˌdeɪt/ v. 巩固；加强
duplicate /'djuːplɪˌkeɪt/ v. 复制；重复

▶ Text understanding

Complete the following table to check your understanding of the major points and structure of the text.

Introduction (Paras. 1-2)	To achieve the common long-term goal for action on climate change, further steps are needed in the COP21 agreement to 1) _____, and a link of 2) _____ is optimal.
Body (Paras. 3-6)	Section 1: Paras. 3-4 **The two steps to lock in the 2°C vision more strongly:** • Re-express that goal in terms of 3) _____. • Create a clear link between 4) _____.
	Section 2: Paras. 5-6 **The effects of the long-term goal on technology development:** • Send a clear message to investors in 5) _____. • Ensure the operation or retirement of 6) _____, allowing only very efficient or low-carbon generating plants to operate. • 7) _____ further energy technology development.
Conclusion (Para. 7)	There are other detailed ways to deal with 8) _____.

Language building

1 Match each of the words in the left column with its corresponding meaning in the right column.

_____	1 valid	a	able to exist or be used together without causing problems
_____	2 boost	b	to make stronger by some action or event
_____	3 optimal	c	a very important part of sth.
_____	4 benchmark	d	standard for measuring or judging other things of the same type
_____	5 trajectory	e	the best or most appropriate
_____	6 alignment	f	reasonable and generally accepted
_____	7 compatible	g	support for a group, political party or country
_____	8 duplicate	h	the way in which a process or event develops over a period of time
_____	9 pillar	i	to help sth. increase, improve or become more successful
_____	10 consolidate	j	to make an exact copy of sth.

2 Among the three choices given, choose the one that is **NOT** close in meaning to the underlined word in each sentence.

1 While this is a valid long-term goal, it is not easy to <u>translate</u> it into practical policy and investment needs.
   A) turn   B) convert   C) write

2 Further steps are needed in the COP21 agreement to lock in the 2°C vision more strongly, so that it <u>anchors</u> future expectations.
   A) secures   B) supports   C) locks in

3 A focus only on short-term emissions targets to 2025 could lead to the adoption of a technology mix that is far from optimal if not <u>devised</u> in the light of a longer-term decarbonization plan.
   A) designed   B) invented   C) discovered

4 A number of possible options for <u>supplementing</u> the 2°C goal in this way have been tabled for consideration in the run-up to COP21.
   A) implementing   B) strengthening   C) complementing

5 The second step is to create a clear link between it and countries' actions, rather than it being just an aspirational <u>statement</u>.
   A) instruction   B) announcement   C) declaration

6 The COP21 agreement must recognize this, <u>embedding</u> climate goals in plans which permit countries to achieve these aspirations.
   A) inputting   B) planting   C) installing

7 The newly expressed long-term goal and low-carbon development pathways will <u>shape</u> decisions on the operation or retirement of existing capital stock and technology development.
   A) influence   B) determine   C) encourage

8 A clear, measurable long-term vision in the COP21 agreement will <u>underpin</u> investment in those key technologies.
   A) strengthen   B) support   C) welcome

9 There are also a number of more detailed ways the COP21 decisions (and subsequent technical work programs) can <u>address</u> aspects of technology development.
   A) deal with   B) cope with   C) stress

10 The technology needs assessments and technology action plans prepared by developing countries should be <u>integrated</u> into wider low-carbon development strategies.
   A) incorporated   B) intersected   C) combined

3 Complete each of the following sentences with an appropriate word from the given word family. Change the form where necessary.

1 The English version is boring – maybe it has lost something in _____. (translate, translation, translator)

2 Have you _____ where you're going for your holiday this year? (decide, decisive, decision)

3 Participation will not be attractive to those whose mortgages already _____ the value of their homes. (exceed, excess, excessive)

4 Britain agreed to cut _____ of nitrogen oxide from power stations. (emit, emission, emissive)

5 She was filled with the _____ to succeed in life. (aspire, aspiration, aspirational)

6 Having a good grasp of the issues and knowing how to _____ are essential when it comes to decision-making. (prior, priority, prioritize)

7 This team came up with a more useful method for _____ new ideas. (generate, general, generative)

8 A sentencing judge is required to consider any _____ circumstances before imposing the death penalty. (mitigate, mitigating, mitigation)

4 Match each word in the box with the group of phrases where it is usually found.

cumulative   unlock   underpin   valid   option   duplicate

1 _____
   a (n) ~ goal
   a (n) ~ reason
   ~ concerns
   a (n) ~ point

2 _____
   possible ~s
   the best ~
   leave your ~s open
   various ~s

3 _____
   ~ emissions
   ~ process
   ~ effect
   ~ loss

4 _____
   ~ long-term emission reductions
   ~ the mysteries
   ~ a secret
   ~ an adventure

5 _____
   ~ investment in those key technologies
   ~ the theory
   ~ your argument
   ~ a statement

6 _____
   ~ these efforts
   ~ previous work
   ~ a document
   ~ other's ideas

5 Find the idiomatic expressions in the text matching the Chinese equivalents.

1 将…转化为 _____
2 关键支柱 _____
3 建立在…基础上 _____
4 过度排放 _____
5 风险评估 _____
6 脱碳计划 _____
7 考虑到; 鉴于 _____
8 绝非; 远不如 _____
9 发展路径 _____
10 振奋之词 _____
11 预算框架 _____
12 修建中 _____
13 将…嵌入 _____
14 审查周期 _____
15 当前资本存量 _____
16 优先考虑 _____
17 与…兼容 _____
18 长期资产 _____
19 低排交通工具 _____
20 效率措施 _____
21 碳捕集与封存 _____
22 新兴技术 _____
23 适应于; 有助于 _____
24 走低的价格 _____

Energy and Climate Change | 157

6 Combine the sentences given below. Then compare your sentences with the original ones in the text.

1 _____

    a Further steps are needed in the COP21 agreement to lock in the 2°C vision more strongly, so that it anchors future expectations, guiding policymaking and both public and private energy sector investment decisions.

    b Further steps are needed in the COP21 agreement to lock in the 2°C vision more strongly, so that it acts as a standard.

    c Short-term government targets and actions can be assessed against the standard.

2 _____

    a A focus only on short-term emissions targets to 2025 could lead to the adoption of a technology mix, resulting in increasing the cost of achieving critical climate goals.

    b The targets themselves are important to avoid excess emissions in the short term.

    c The technology mix is far from optimal if not devised in the light of a longer-term decarbonization plan.

3 _____

    a There are a number of possible options for supplementing the 2°C goal in this way.

    b These options have been tabled for consideration in the run-up to COP21.

    c These options include a target level for 2050 emissions, a target date for net-zero emissions in the second half of this century, or a timeline for the phaseout of unabated fossil fuels.

4 _____

    a The global goal should be used by all countries to inform nationally determined decarbonization or low-carbon development pathways.

    b Nationally determined decarbonization or low-carbon development pathways would in turn provide an important benchmark for the short-term goals.

    c Countries set the short-term goals in the five-year review cycle.

5 _____

   a A clear, measurable long-term vision in the COP21 agreement will underpin investment in those key technologies.

   b Those key technologies are essential to unlocking long-term emission reductions.

   c The long-term emission reductions should be compatible with a 2°C trajectory.

6 _____

   a The technology needs assessments and technology action plans are prepared by developing countries.

   b These assessments and plans should be integrated into wider low-carbon development strategies.

   c The integration improves alignment between these countries' development, mitigation and technology goals.

7 Translate each of the Chinese sentences into English by using the underlined phrase or structure in the example.

1 While this is a valid long-term goal, it is not easy to <u>translate</u> it <u>into</u> practical policy and investment needs.
他在大学里学到的工作方法不适合商业界。
_____
_____

2 It <u>acts as</u> a standard against which short-term government targets and actions can be assessed.
联合国将作为和平解决方案的保证人。
_____
_____

3 ... could lead to the adoption of a technology mix that is far from optimal if not devised <u>in the light of</u> a longer-term decarbonization plan, resulting in increasing the cost of achieving critical climate goals.
考虑到她突发感冒，约会取消了。
_____
_____

4 This would be more straightforward to apply in the energy sector and <u>lend itself to</u> easier accounting for monitoring.
她的嗓子不是很适合唱布鲁斯歌曲。
_____
_____

5   A trajectory <u>compatible with</u> a 2°C objective that fully internalizes a long-term decarbonization vision would need to cut cumulative emissions by around 35%.
    他认为自由企业制并不违背俄罗斯的价值观和传统。
    _____

6   Care should be taken to <u>build on</u>, consolidate and not duplicate these efforts.
    这项研究是在以往工作的基础上进行的。
    _____

7   A number of possible options for supplementing the 2°C goal in this way have been tabled for consideration in <u>the run-up to</u> COP21.
    书出版前的最后准备阶段，每个人都在紧张地忙碌着。
    _____

Academic writing

▶ Micro-skill: Argument and discussion

Academic writing often involves arguments that are debatable. This means such an argument needs to be discussed from both sides of the argument before a conclusion is made in a suitably academic manner.

1 Structures of discussion

There are two structures of discussion. The vertical structure may be made up of a paragraph of advantages and a paragraph of disadvantages. The horizontal structure is developed by examining the topic from different viewpoints.

Discussion: It is a common practice all over the world to send criminals to prison and this is considered as the best way to deal with them. However, some people suggest that alternative measures are more effective. So should all criminals be put into prison?

Vertical

Advantages:
1) It would have a deterrent effect on those who intend to commit crimes.
2) In prison, criminals can be properly educated, treated and fundamentally reformed.
3) It is fair for victims and their relatives to see the criminals imprisoned.

Disadvantages:
1) When there are too many criminals, more prisons should be built.
2) Instead of being reformed, some criminals may learn from some other criminals and will probably re-offend after leaving.
3) When criminals are confined in prison for a long time, they may suffer from some mental problems.

Horizontal

Economic: If all criminals were put into prison, it would be expensive to build prisons.

Ethical: Considerations should be given to the rights prisoners should have.

Social: If criminals fail to master the basic skills of making a living or do not have a proper legal sense, they are more likely to commit crimes again. However, some violent criminals should always be put into prison.

1 Brainstorm the positive and negative aspects on whether there should be more multinational enterprises, and then write an outline using one of the structures (vertical or horizontal) above.

Positive	Negative
• The trend of globalization gives a boost to the world economy. • _____ • _____	• In developing countries, a large number of local companies closed down, as they were defeated in the competition with their transnational counterparts. • _____ • _____

Outline:

2 Write two paragraphs on the topic "Are Robots Essential for Human Beings in the Future?". Use the ideas below and make your stance clear.

Pros
- The various applications of robots
- How human beings can benefit from the application of robots
- The important role robots play in the development of human beings

(Source: Donnet-Kammel, 2005)

Cons
- Robots can cause troubles and confusion for human beings.
- Robots can be misapplied or fail to work properly.
- It is dangerous when robots are out of human control.

(Source: Soroka, 2000)

2 Language for discussion of an argument

The following expressions can be used in discussing an argument.

Positive	Negative
benefit	drawback
advantage	disadvantage
a positive aspect	a negative feature
pro (informal)	con (informal)
plus (informal)	minus (informal)
one major advantage is …	a serious drawback is …

Examples:

One drawback to this method is the time and expense that are required to implement it.

A significant benefit of the program is the increased income of all workers.

3 Complete the following paragraph using expressions from the table above.

There is an increasing number of students going abroad to study. Studying abroad can have significant 1) _____, such as the opportunity to understand diversified cultures. But it also involves some 2) _____, like feelings of isolation or homesickness. Another 3) _____ side may be the high cost, including both fees and living expenses. However, most students are in the view that the 4) _____ outweigh the 5) _____, and that the opportunity to be a part of an international group of students is a major 6) _____ in terms of their career development.

▶ Macro-skill: Conclusion

1 Overview

The end of an essay is important because it is this part that gives the reader the final, and often the deepest impression. Not every essay needs a separate concluding paragraph. For a short essay, the last paragraph of the main body, even the last sentence of that paragraph may serve as the conclusion so long as it gives the reader a sense of wholeness. But more often, an independent paragraph is required to conclude a relatively long complete essay. The concluding paragraph of an essay summarizes the main points to help the reader remember them.

Since the concluding paragraph (also called closing or ending paragraph) is the last part of the essay, it takes a lot of time and energy to make it as forceful as possible, made up mainly of restatements or a summary of the points that have been discussed in the main body.

2 Ways of concluding an essay

Here are some strategies for writing an effective ending / concluding paragraph.

1) Concluding with a restatement

Probably the most common type of ending is one which restates the controlling idea raised in the introductory paragraph, better in different words. If, for instance, a question is raised in the introductory paragraph, a definite answer should be given in the concluding paragraph.

<u>The introductory paragraph:</u>

Views on whether we should learn classics or not vary from person to person. Some hold that classics, having stood the test of time, should be treasured and learned by all generations of people while others argue that these classics are out of date and not worth our time and energy on them. In my opinion, however, learning classics is very important.

<u>The concluding paragraph:</u>

In summary, not only do classics record our past, but also they remain relevant to our present, even future. Learning classics is still important and worthwhile today because the way and wisdom reflected in these classics are not only the record of our ancestors, but also can lead our life today.

2) Concluding with a generalization

The introductory paragraph:

Studies of implicit bias have recently drawn anger from both the right and the left. For the right, talk of implicit bias is just another instance of progressives seeing injustice under every bush. For the left, implicit bias diverts attention from more damaging instances of explicit prejudices. Debates have become heated, ranging from scientific journals to the popular media. Along the way, some important points have been lost. We highlight two misunderstandings that anyone who wants to understand implicit bias should know about.

This introductory paragraph is followed by a detailed analysis of the "two misunderstandings" from various perspectives, after which a concluding / generalizing paragraph finalizes the whole passage.

The concluding paragraph:

One reason people on both the right and the left are skeptical of implicit bias might be pretty simple: It isn't nice to think we aren't very nice. It would be comforting to conclude, when we don't consciously entertain impure intentions, that all of our intentions are pure. Unfortunately, we can't conclude that: Many of us are more biased than we realize. And that is an important cause of injustice – whether you know it or not.

3) Concluding with a quotation or anecdote that indicates or reinforces the central idea

The introductory paragraph:

Lower salaries aren't the only factor holding women back from accumulating wealth. The gender wage gap gets a lot of attention, and for good reason. At the current snail's pace by which it's closing worldwide, the World Economic Forum says it will take about 200 years to close the pay gap. But there are other financial imbalances holding many women back and keeping them from economic independence.

As has been learned in Unit 1, "but" is often used to introduce the writer's idea or topic for discussion. Then we can predict that this introductory paragraph is followed by a list of "other financial imbalances" besides the gender wage gap. Actually five such "imbalances" are included in the main body of the passage, respectively, "the retirement savings gap", "the student debt gap", "the financial literacy gap", "the work time gap" and "the homeownership gap".

The concluding paragraph:

Finance experts agree that investing in closing these gaps will benefit Americans as a whole. "White women are still earning only 80 cents on the dollar of their male counterparts. It is even worse for women of color," says Rebecca Wiggins, Executive Director of AFCPE, a nonprofit training finance professionals to understand the gap. "This decreased earning potential impacts things like compounding interest which leads to wealth inequality.

"Women are becoming the primary or sole breadwinners of their families and often make the purchasing decisions for the household, so there are significant implications for the growth and stability of our national economy by closing the gender wealth gap."

4) **Concluding with a call for action or suggestion**

The first two paragraphs:

Both distance-learning courses and traditional classes provide important but different experiences for college students. On the one hand, there are many advantages of distance-learning courses. One of the most important benefits is the opportunity to attend class at your own convenience. This is very important for students who have full-time jobs since they can choose to take classes on a schedule that allows them to continue working. Another advantage is the chance to complete assignments at your own pace. For students who work more quickly than their classmates, it is possible to earn more credits during a semester. A huge advantage is that students can listen to the lectures more than once.

On the other hand, there are advantages of attending a traditional class. The structured environment is beneficial, especially for students who are not highly motivated. In addition, it is more likely that you will develop a personal relationship with the lecturer / instructor, an advantage not only for the course but also after the course when you need a recommendation. By seeing you and talking to you face-to-face, the teacher will remember you better. It is also easier to get an immediate response to questions because you only have to raise your hand instead of sending an e-mail and waiting for an answer. Last but not least, the opportunity for study groups and friendships is different and more personal when you sit in the same room.

The concluding paragraph:

Given all the advantages of both types of courses, it would be wise to register for both distance-learning courses and traditional classroom courses during their college life. By participating in the former, they can work independently in courses that may be more difficult for them, repeating the lectures on computer at convenient times. By attending traditional classes, they can get to know the teachers personally and will have good references when necessary. They can also make friends in the class. By sharing information with other students, they can organize their schedules for the following semester, choosing the best classes and including both distance-learning and traditional courses.

Read the following passage and complete the table.

In 2016 the video gaming industry racked up sales of about $100 billion, making it one of the world's largest entertainment industries. The games range from time-wasting smartphone apps to immersive fantasy worlds in which players can get lost for days or weeks. Indeed, the engrossing nature of games is itself cause for concern. Last year four economists published a paper suggesting that high-quality video games – an example of what they call "leisure luxuries" – are contributing to a decline in work among young people, and especially young men. Given the social and economic importance of early adulthood, such a trend could spell big trouble. But are video games causing the young to turn on and drop out?

In making the link between gaming and work, the economists – Mark Aguiar, Mark Bils, Kerwin Charles and Erik Hurst – point to compelling data. From 2000 and 2015, the employment rate for men in their 20s without a college education dropped 10 percentage points, from 82% to 72%. Such men often live at their parents' homes and tend not to marry at the same rate as their peers. They, do, on the other hand, play video games. For each hour less the group spent in work, time spent at leisure activities rose about an hour, and 75% of the increased leisure time was accounted for by gaming. Over the same period games became far more graphically and narratively complex, more social and, relative to other luxury items, more affordable. It would not be surprising if the satisfaction provided by such games kept some people from pursuing careers as aggressively as they otherwise might (or at all).

To draw a firm conclusion, however, would take a clearer understanding of the direction of causation. While games improved since the turn of the century, labor-market options for young people got worse. Hourly wages, adjusted for inflation, have stagnated for young college graduates since the 1990s, while pay for new high-school graduates has declined. The share of young high-school and college graduates not in work or education has risen; in 2014 about 11% of college graduates were apparently idle, compared with 9% in 2004 and 8% in 1994. The share of recent college graduates working in jobs which did not require a college degree rose from just over 30% in the early 2000s to nearly 45% a decade later. And the financial crisis and recession fell harder on young people than on the population as a whole. For people unable to find demanding, full-time work (or any work at all) gaming is often a way to spend some of one's unwanted downtime, rather than a lure out of work; it is much more a symptom of other economic ills than a cause.

Games will go on getting better, and the share of jobless or underemployed young people choosing to game rather than focus on career will probably grow. That is not necessarily something to lament. Games are often rewarding and social, and time spent gaming sometimes displaces less healthy or rewarding pastimes. If the pull of work is not strong enough to overcome the desire to game, the first response should be to ask whether more can be done to prepare young people for good jobs – and to make sure that there are some around when those young people enter the workforce.

Topic	The question raised for discussion: 1) _____?

Analysis
(Facts and causes)
	Facts (compelling data): 2) _____.
	Causes of the phenomena:
	3) _____.
	   a. _____.
	   b. _____.
	   c. _____.
	   d. _____.
	Summary: 4) _____.

Conclusion	5) _____. 6) _____.

Writing assignment

Write a concluding paragraph for the writing assignment in Unit 1. You should list the main points of the passage before composing the conclusion. The concluding paragraph should be around 120 words.

1 Energy is indispensable in modern society.

2 Energy is the engine of civilization.

Sharing

> You are the office secretary of the municipal government, and now you are assigned the task of composing an energy development plan for the city against the background of global climate change. You are asked to present the energy development plan in the hearing meeting.

1 Work in groups to collect relevant information and statistics both in the texts and from the Internet. The information includes the relationship between energy consumption and climate change, regulation of the energy industry to reduce carbon emissions, and ways to encourage the development of clean energy.

2 Summarize and integrate the information that you have collected. Try to set a working agenda on how to develop local economy in a way that will not contribute to global climate change.

3 Report your energy development plan in class. Your presentation should include the following parts:

1) Relationship between energy consumption and climate change
2) Measures to regulate the local energy industry to reduce carbon emissions
3) Measures to encourage the development of clean energy
4) An integrated working agenda of the measures above